La Jolla 2008

Reflections on Time and Politics

Reflections on Time and Politics

Nathan Widder

The Pennsylvania State University Press
University Park, Pennyslvania

Library of Congress Cataloging-in-Publication Data

Widder, Nathan, 1970–
Reflections on time and politics / Nathan Widder.
p.   cm.
Summary: "Explores the nature of time and its implications for questions of politics, ethics,
and the self. Shows how a conception of time that breaks with common sense notions of
chronological order can help us rethink the understandings of identity, difference, power,
resistance, and overcoming"—Provided by publisher.
Includes bibliographical references (p.   ) and index.
ISBN 978-0-271-03394-5 (cloth : alk. paper)
1. Time.
I. Title.

BD638.W53 2008
115—dc22
2008007352

IN MEMORY OF MY FATHER, ROBERT WIDDER
(Oct. 23, 1926–Feb. 21, 2008),
*who faced everything that came his way
with courage and dignity.*

In our failure to understand the use of a word we take it as the expression of a queer *process*. (As we think of time as a queer medium . . .)
—WITTGENSTEIN, *Remarks on the Foundations of Mathematics*, 1.127

We would know nothing of time and motion if we did not, in a coarse fashion, believe we see what is at "rest" beside what is in motion.
—NIETZSCHE, *The Will to Power*, §520

Philosophy's sole aim is to become worthy of the event.
—DELEUZE AND GUATTARI, *What Is Philosophy?* p. 160

CONTENTS

PREFACE AND ACKNOWLEDGMENTS

I BEGAN THIS PROJECT in early 2004, shortly after a five-month visit by Professors William E. Connolly and Jane Bennett to my department at Exeter University. I benefited greatly from the papers they delivered and from our many lengthy discussions. Bill, who came to Exeter as a Leverhulme Visiting Professor, also led an engaging "Time and Politics" research seminar, which gave me the opportunity to carry out my first in-depth reading of Bergson's *Creative Evolution* and to present a paper on Deleuze's three syntheses of time, which has developed into the center of this book. At the time I was working—or at least claiming to be working—on a book-length study of Deleuze. However, despite the breadth of Deleuze's thought and my strong intellectual debt to it, the focus on a single thinker was becoming increasingly unsuitable for the diversity of the avenues I wanted to explore. This diversity, which now included a number of issues around the nature of time and its relations to politics and ethics, also made the standard book format of five or six chapters undesirable. I wanted focused components that were shorter than chapters but more sustained than aphorisms, and that could be loosely organized as a series of explorations of or engagements with a broad range of topics and thinkers. After a period of reflection on my experiences in Bill's seminar and how I could develop the thoughts it inspired in me, the idea for this project and its format came to me. Hence the title: *Reflections on Time and Politics*.

This project combines the interest in time that developed as a result of Bill and Jane's visit with a long-felt dissatisfaction with contemporary political philosophies that see identity being formed through constitutive exclusions. To my mind, these accounts end up holding the consolidation of identity to be contingent and ephemeral, yet also to be the *sine qua non* for political thought and action. While this approach to identity is understandable and, in certain areas of politics, often useful and effective, I have long found it limited. This has led me to ideas associated with Deleuze and Guattari's micropolitics and Foucault's care for the self, which seem to me to go in a very different direction from most other postidentity political theory. It has also led me to the thesis from Deleuze's late 1960s writings that holds identity to be a simulation or optical illusion. I am well aware that Deleuze moved away from this terminology. Nevertheless, it is appropriate to the thesis I will be advancing here: that identity and fixed markers, which may be considered natural and pregiven or contingently constructed but indispensable, are surface effects of difference. Identities and fixed markers, I want to say, are like patterns on the surface of water, which appear fixed when seen from a great distance, such as from the window of an airplane in flight: their stability and substantiality, in short, are a matter of perspective.

As the pages that follow will try to demonstrate, to hold that identities are semblances of stability is not to suggest that they are unimportant or dispensable. Indeed, they structure a great deal of personal, social, and political life. Nor is it to say that identity can be easily modified. Just as a shadow on a wall, being cast by an object located elsewhere, cannot be changed by attempting to alter the image directly, so identity must be modified by adjusting the flowing relations of difference that project it as a fixed center. And the case of identity is more complicated because its source is not a solid object, making the task, again using the water metaphor, more akin to shifting currents that can easily slip around barriers, return to their original paths once these barriers are removed, and produce similar patterns even when they are successfully altered. That identities and fixed markers must be approached indirectly through the often hidden and subtle dynamics that constitute them is what makes micropolitics and practices of the self so important; that these same identities appear to be more significant than they are is what makes the undertaking so challenging. Identity is certainly characterized by endurance over time, but it is an error, I would argue, to equate endurance with substantiality. Someone may have a seemingly stable personality characterized by generosity, a sarcastic sense of humor, and a short temper. However, since no one remains the same either materially (our bodies constantly change) or "spiritually" (ever more layers of our past embed themselves in us, our comportment toward the future that approaches us shifts in different ways, we have changing relationships with different people, and so on), it must be the case that if someone remains generous, sarcastic, and short tempered over time, it cannot be for the same reasons, because he or she is not even the same person. The constants are only the somewhat regular results of the varying "syntheses" of ever-changing constitutive relations that make us who we are at particular moments and in particular contexts. Their consistency, which certainly exists, is in the effect, not in the being, making them in this sense extra-being and surface phenomenon. Only a Platonist, it seems to me, would find this automatically condemnable. Foucault uses similar terms in *The Archaeology of Knowledge* when he holds a discursive formation to be not a static form but a schema of correspondence between flowing temporal series that gives sense to the subjects and objects that, whether ephemeral or enduring, are underpinned by the formation. So too does Deleuze in *Difference and Repetition,* when he writes of identity revolving around the different. I feel there is a political and ethical point to viewing personal and collective identity and related markers that organize personal, social, and political life in this way. This is what this book tries to develop, using the structure of time as its ontological linchpin.

Several parts of this work are based on previously published articles and chapters. Although they have been updated and rewritten to different degrees, they nevertheless largely reproduce these earlier publications. I would like to thank the various editors and publishers for their permission to incorporate material from the following: "Thought After Dialectics: Deleuze's Ontology of Sense," *Southern Journal of Philosophy* 41, no. 3 (2003): 451–76; "Foucault and Power Revisited," *European Journal of Political Theory* 3, no. 4 (2004): 411–32; "The Relevance of Nietzsche to Democratic Theory: Micropolitics and the Affirmation of Difference," *Contemporary Political Theory* 3, no. 2 (2004): 188–211 (reproduced with permission of Palgrave Macmillan, Ltd.); "Two Routes from Hegel," in *Radical Democracy: Politics Between Abundance and Lack,* ed. Lars Tønder and Lasse Thomassen (Manchester: Manchester University Press, 2005), 32–49; and "Time Is Out of Joint—and So Are We: Deleuzean Immanence and the Fractured Self," *Philosophy Today* 50, no. 4 (2006): 405–17.

This project also benefited from fellow scholars who offered valuable advice and encouragement or who simply shared the same intellectual space with me. In many instances they also read and commented on draft versions of some or all of this work. In addition to the already mentioned William Connolly and Jane Bennett, I would like to thank Keith Ansell Pearson, Terrell Carver, Dario Castiglione, Diana Coole, Robin Durie, Johnny Golding, Matthew Hammond, Iain Hampsher-Monk, Ed Kazarian, Will Large, Len Lawlor, Iain MacKenzie, Paul Patton, Jon Simons, Daniel W. Smith, Lasse Thomassen, Lars Tønder, Yves Winter, Shane Wolfland, and Martin Wood.

Finally, this work was largely composed during a period of great personal difficulty, and so I would like to thank all of the family members, colleagues, and friends whose unwavering support helped me through this time.

Introduction

RECENT PHILOSOPHICAL DEBATES have moved beyond proclamations of the "death of philosophy" and the "death of the subject" to consider more affirmatively how philosophy can be practiced and how the human self or subject can be conceptualized today. Combined with the impact of profound changes related to globalization and the information age economy, which have blurred both real and conceptual boundaries and made speed a central factor of contemporary life, this has produced a renewed interest in time as an active force of change, contingency, and novelty. Time's dynamics, embedding the past and memory in the present in such a way as to propel time into an always open and indeterminate future, are being deployed against spatial models of representation to develop further nonrepresentational concepts of difference, which had often been elaborated primarily in spatial terms. Time's multiplicity, manifesting itself in the coexistence of different tempos and velocities of time, is now used to account for complex and ambiguous processes of contemporary life and to outline excessive forms of speed that modern societies seek to control. These developments are also creating new connections between philosophy and life sciences—particularly evolutionary theory and neuroscience—that are proving invaluable in analyzing the emergence of complex and ever-changing systems. The once-forgotten Bergson is a central figure in these new political philosophies of time, as is Deleuze, whose increasing prominence on the Anglo-American scene has driven much of this Bergson revival. As with many critiques of representation, the goal of these explorations into time, complexity, and politics is the development of a pluralist politics and ethics.[1]

1. Recent works of pluralist philosophy that engage with these issues and have been inspired both by Deleuze and Bergson and by evolutionary theory and neuroscience include Ansell Pearson (1999 and 2002), Connolly (2002), and Grosz (2004). Although not particularly indebted to Bergson, Bennett (2001) also deserves mention.

This often explicit focus on time's *movement*, however, seems problematic.[2] Perpetual movement is certainly time's first and most obvious trait; as Bergson says, "the essence of time is that it goes by" (Bergson 1991, 137). Moreover, emphasizing the dynamism of both time and the processes occurring in it has obvious advantages for a pluralist political thought that aims to challenge conventional identity-based politics by demonstrating both the contingent, incomplete, and fluctuating nature of identity and its slippery materiality.[3] Finally, stressing time's movement helps foster a more natural link between philosophy and the physical sciences, whether the latter are appropriated positively to develop more complex understandings of social and cultural life or subjected to Bergsonian critiques of holding abstract conceptions of time inappropriate to the study of concrete human existence and its duration. Privileging time's passage, however, seems to me to be both analytically incomplete and inconsistent with respect to the principal philosophical sources inspiring this move. It is not at all clear, when considering these sources in detail, whether they share this privileging of time's movement, whether it is justified, and whether it can ground the innovation that current philosophies of time promise. For example, does not Bergson's *Creative Evolution*, despite characterizing concrete time as continual variation, provide only a negative explanation for its positive creativity, an explanation by default?[4] In contrast, does Deleuze not demand a positive principle to account for time's constitutive repetitions?[5] Although Deleuze correctly notes that Bergson goes beyond the flowing image of time

2. Grosz, for example, declares that she "is concerned primarily with the question of the ontology of time, duration, or becoming, the ontological implications for living beings of their immersion in the always forward movement of time" (2004, 4).

3. The vitalism and focus on becoming and materiality that have been associated with Deleuze have made him an inspiration for recent theories of "radical democracy" grounded in "ontologies of abundance." See Tønder and Thomassen (2005).

4. Time's power cannot be merely destructive, as this is easily susceptible to mechanistic explanation (Bergson 1998, 19). Instead, Bergson maintains, the gnawing of the past into the present creates an irreducibly new future (21–22). Yet this claim, aside from being presented through the metaphorical language of *élan vital*, is based primarily on the incompleteness of mechanism and finalism: "As soon as we go out of the encasings in which radical mechanism and radical finalism confine our thought, reality appears as a ceaseless upspringing of something new" (46–47); also, "just because it goes beyond intellect . . . this reality is undoubtedly creative, *i. e.* productive of effects in which it expands and transcends its own being" (52). At the same time, however, while mechanism is disproved by "the slightest trace of spontaneity" (40), Bergson must admit that "the doctrine of final causes . . . will never be definitively refuted" (40) and that creative evolution will always retain a proximity to finalism (50).

5. "Our problem concerns the essence of repetition. It is a question of knowing why repetition cannot be explained by the form of identity in concepts or representations; in what sense it demands a superior 'positive' principle" (Deleuze 1994, 19).

to present a static transcendental image—that of the coexistence of virtual past and actual present—does not Deleuze break with Bergson precisely over this image, holding that it merely grounds the moving image of time, reinstates a moment of transcendence, and fails to account for creativity, particularly the creativity of thought? Current philosophies of time often treat Bergson as Deleuze's chief inspiration. Yet is Bergson's insistence on time's continuity not fundamentally incompatible with Deleuze's idea of time as a "disjunctive synthesis"?[6] The easy association of Bergson and Deleuze seems to depend on the assumption that time's flow is their shared ultimate concern and that it is sufficient to explain time's production of novelty.[7]

It is not that time's flow is unimportant, but it is neither primary nor foundational. In many respects focusing on time's passage or on the static conditions that make this passage possible continues time's long-established subordination to movement and space. Therefore, despite the important advances the recent interest in time has brought to pluralist political philosophy, this work sees the need to take a further step back and consider time as a static structure or synthesis that *ungrounds* movement, including the movement of time. This decision is inspired largely but not exclusively by Deleuze, who consistently executes the same maneuver, from his early analysis of the syntheses of time to his late work on the time-image in cinema (Deleuze 1989, 1994). The aim in following Deleuze here is to further rethink the nature of both time and change in terms that I feel have not been considered sufficiently in contemporary political thought. These terms are also inspired primarily by Deleuze and particularly by a thesis that is articulated in his earlier work, but whose significance remains central, despite certain renunciations, in his later writings: the thesis that identity is a simulation or optical illusion.[8] This idea is not attained simply by

6. While Deleuze's writings (particularly 1991 and 1999) certainly draw a conception of immanent difference from Bergson, this difference cannot be properly equated with the ontological difference or differenciator of Deleuze's disjunctive synthesis. Indeed, as will be seen, Deleuze provides an unconvincing argument to bring Bergson to this level before ultimately leaving Bergson behind and turning to Nietzsche (particularly in Deleuze 1994).

7. Deleuze's break with Bergson, along with many recent interpretations that downplay their fundamental differences, will be addressed primarily in the fourth and eighth reflections of this work. For now, one should recall the whiff of finalism in Bergson's account of the evolution of the human consciousness, intellect, and freedom (see Bergson 1998, 176–85).

8. In his "Letter-Preface" to Jean-Clet Martin's 1990 work, *Variations—La philosophie de Gilles Deleuze*, Deleuze writes: "Similarly, you grasp the importance I assign to the notion of multiplicity: it is essential. As you say, multiplicity and singularity are intimately connected ('singularity' being at once different from 'universal' and 'individual'). 'Rhizome' is the best

rejecting all substrata beneath change and thereby submitting everything to motion, as a center of continual change and novelty can still be retained. Thus Bergson, even while rejecting the "formless *ego*, indifferent and unchangeable, on which it [the attention] threads the psychic states which it has set up as independent entities" (Bergson 1998, 3), suggests no qualms with an ego that both changes and endures (1998, 4; for an explicit acceptance of the idea of the concrete ego, see Bergson 1910, 226; 1983, 71).[9] Instead, the idea that identity—along with its associated conceptions of difference, stability, and endurance—is a simulation comes to light by considering not only time's motion and the structure grounding it, but more deeply time's structural ungrounding. This ungrounding, in turn, shifts the nature of both time and movement.

This work will argue that this consideration of structural ungrounding and the concomitant status of identity as a simulation have profound implications for the entire way in which power, meaning, and resistance—as well as time and change—are theorized for pluralist politics and ethics. The move from ontology to politics and ethics often seems precarious, yet this apparent difficulty is also reminiscent of the sort of false problems Bergson attributes to most philosophical endeavors. Insofar as time is fundamental

term to designate multiplicities. On the other hand, it seems to me that I have totally abandoned the notion of simulacrum, which is all but worthless. *A Thousand Plateaus* is the book dedicated to multiplicities for themselves (becomings, lines, etc.)" (Deleuze 2006, 362). Deleuze certainly has good tactical reasons for relinquishing the language of simulacra and the simulacrum, given how they are frequently reduced to their most generic meanings as referring to a copy without original or a difference without real status. On the one hand, these meanings had underpinned claims prominent in social and cultural analysis about reality melting into a postmodern hyperreality of simulation (see, for example, Baudrillard 1993, chapter 2) long before Deleuze and Guattari wrote: "Philosophy has not remained unaffected by the general movement that replaced Critique with sales promotion. The simulacrum, the simulation of a packet of noodles, has become the true concept; and the one who packages the product, commodity, or work of art has become the philosopher, conceptual persona, or artist" (1994, 10). On the other hand, several critical attacks attribute these meanings to Deleuze (most infamously Badiou 2000; in reply to Badiou, see Widder 2001). Nevertheless, Deleuze never abandons the two central ideas that he initially designated with these terms: that of a structure of differences not modeled on any identity and relating differences through their difference (the structure of the rhizome or multiplicity), and that of the transcendental illusion of a center or identity generated by the dynamics of this decentered system ("the plane [of immanence] is surrounded by illusions. These are not abstract misinterpretations or just external pressures but rather thought's mirages" [Deleuze and Guattari 1994, 49]). These are the meanings designated by simulacrum and simulation throughout this work.

9. Sartre maintains that Bergson's reference to the ego in *Time and Free Will* restricts the spontaneity of consciousness because Bergson fails to recognize "that he was describing an *object* rather than a consciousness, and that the union posited is perfectly irrational because the producer is passive with respect to the created thing" (Sartre 1957, 80).

to human experience, an ontology of time is a human (although not a humanist) ontology. The structural ungrounding of time, as will be seen, is part and parcel of the being of language, the unconscious, memory, and thought. The ontological stakes involved in this rethinking of time—particularly those involving difference, immanence, surface sense, and the being of the simulacrum—therefore speak directly to the political and ethical stakes that are this work's ultimate concern.

This project thus aims to examine the philosophical, ethical, and political implications of an ontology that releases time from its traditional subordination to movement. According to an ordinary or common conception—whose philosophical counterpart is frequently attributed, rather problematically, as will be seen, to Aristotle—"time" names an experience of passage that can serve to measure rates of change and movement. It is a chronological succession, modeled on the picture, according to Bergson and Wittgenstein, of the film strip that ceaselessly passes and is counted off to calculate change.[10] It is well known that this approach to time creates irreconcilable aporias within traditional metaphysical categories, a fact that has led Kantian and post-Kantian philosophy to explore transcendental conditions of constitution. The results, such as Heidegger's thesis that temporality is the horizon of being[11] and Deleuze's declaration that repetition in the eternal return is the "for-itself" of difference (Deleuze 1994, 125), invert the standard priority, so that instead of time being modeled on movement, movement is modeled on it. Wittgenstein sums up the difference:

> The feeling we have is that the present disappears into the past without our being able to prevent it. And here we are obviously using the picture of a film strip remorselessly moving past us, that we are unable to stop. But it is of course just as clear that the picture is misapplied: that we cannot say "Time flows" if by time we mean the possibility of change. What we are looking at here is really the possibility of motion: and so the logical form of motion. (Wittgenstein 1975, §52)

If time itself moved, Deleuze argues, it would imply another time within which it flowed, and so on ad infinitum; to avoid this infinite regress, time

10. See Bergson (1998, chapter 4) for his most extended discussion of the "cinematographical" conception of time. For Wittgenstein's discussion, see 1975 (§§47–56).

11. "The term 'Temporality' is intended to indicate that temporality, in existential analytic, represents the horizon from which we understand being" (Heidegger 1982, 228).

must be conceived as unchanging form (see Deleuze 1984, vii–viii).[12] Considered on these terms, "time" names the structure, not the measure, of change. It is a kind of *being out of sync with oneself* that is the condition for anything to change or move.

This reversal of time and change, however, carries certain idealist risks. Adorno calls it the danger of the "detemporalization of time," which, removing time from its flux, turns it into an empty form. "When Kant turns time, as the pure visual form and premise of everything temporal, into an a priori, time on its part is exempted from time. Subjective and objective idealism concur in this, for the basic stratum of both is the subject as a concept, devoid of its temporal content. Once again, as to Aristotle, the *actus purus* becomes that which does not move" (Adorno 1995, 331–32). A time cut off from any possible content in this way "would no longer be what time, according to Kant, must inalienably be: it would no longer be dynamic" (332). Although Hegel continues this detemporalization by privileging the subjective and universal in his philosophy of history (331) and deriving time from logic (333), Adorno sees a possible advance suggested by dialectical synthesis and sublation, wherein "a relationship of form and content has become the form itself" (333). This would be a relationship in which form is its own content, making it a self-expression or expression of *sense*. Considered as such a synthetic form, time would not be empty; rather, being the structure of movement or change, it would express the sense of becoming as such.

As the sense and structure of becoming, then, time is a synthesis of differences. There is, of course, a synthesis carried out by consciousness that establishes a linear order of past, present, and future, without which change could not be perceived. But there are also more profound levels of synthesis in relation to which the time of consciousness is only an epiphenomenon. Time here becomes truly creative, and it is not surprising that both Bergson and Nietzsche challenge mechanistic causality and determinism with understandings of time as a nonlinear synthesis occurring underneath the realm of consciousness. With Bergson's duration, the coexistence

---

12. McTaggart's seminal article (1908) employs this reasoning to argue against the reality of time: time cannot be defined without reference to passage—the dynamic A-series of past, present, and future—but this passage either presupposes another time series in which it moves or entails a vicious circle where it is defined by reference to its own terms. For McTaggart, this leaves open the possible reality of a nontemporal C-series as the static order of events that humans experience as passing "in time." Of course, the argument goes back at least to Kant: "For change does not affect time itself, but only appearances in time. . . . If we ascribe succession to time itself, we must think yet another time, in which the sequence would be possible" (Kant 1965, A183/B226)

of the virtual past in the actual present is the precondition for change, creation, and creation, and continuity; its absence would leave only the abstract and artificially closed world of mechanism (see Bergson 1998, chapter 1). Nietzsche too insists against mechanistic theory that "becoming drags the has-been along behind it" (Nietzsche 1982, §49). However, the thought of eternal return, which arises from but also overturns mechanism (Nietzsche 1968, §1066), rejects Bergson's thesis of continuity. If duration expresses an inherence of the past in the present that grounds change, the eternal return expresses the being out of sync with oneself that ungrounds—or serves as the groundless ground—of change. One version of the eternal return holds that, given an infinity of time, all events must repeat themselves endlessly. Nietzsche, however, introduces this as merely an extreme idea, which follows the opposite and equally fallacious extreme of a teleological worldview ("Extreme positions are not succeeded by moderate ones but by extreme positions of the opposite kind" [§55]).[13] As Deleuze argues—and here he consolidates his break with Bergson—the eternal return concerns the way time and things existing in it are structured discontinuously, so that what "returns" is never identity or sameness, but only difference and divergence.

The inversion of time and movement relates to another contemporary philosophical project: the achievement of immanence. This endeavor, aspiring to surpass the standard metaphysical dichotomies of essence/appearance, subject/object, and so forth, is compelled toward a synthesis of differences that would establish the internal passage from one side of these divides to the other without recourse to a transcendent identity, telos, or Form. An ontology of immanence, Deleuze says, is thus one of sense rather than (transcendent) essence (Deleuze 1997b) and it establishes a *surface* that separates and brings together differences, allowing nothing to transcend it by escaping to the heights or falling into the abyss. Time, as the structure of change, would be this very synthesis. Hegelian dialectics also promises

---

13. When Nietzsche presents the eternal return as a recurrence of the same, he bases his view on the presuppositions that "the world may be thought of as a certain definite quantity of force and as a certain definite number of centers of force—and every other representation remains indefinite and therefore useless." Mechanistic theories share these assumptions, yet the eternal return "is not simply a mechanistic conception," as it goes beyond mechanism's idea of a final state of becoming, exposing mechanism as "an imperfect and merely provisional hypothesis" (1968, §1066). However, once this realization is made, the underlying assumption of a fixed quantity of centers also falls: when centers are conceived as "only dynamic quanta" (§635) and mechanism's center of force, the atom, is dismissed as incoherent (see §§624–25, 634–35), it follows that there is, "instead of 'cause and effect' the mutual struggle of that which becomes, often with the absorption of one's opponent; the number of becoming elements not constant" (§617).

immanence through synthesis, using the movement of negation to reconcile differences by both specifying the identity of a thing in relation to what it "is not" and establishing an internal unity between these opposites. Dialectical movement thus progresses toward immanence through the power of an ontological difference—speculative contradiction—that incorporates those differences that initially appeared transcendent. However, as will be seen, this synthesis completes immanence only abstractly. Abstraction here should be taken in Hegelian terms, signifying what is isolated and one-sided: dialectical synthesis remains incomplete because it refers to differences it must exclude or leave unsynthesized, so that in its internal movement of negation and sublation, crucial gaps remain, particularly between the logical structure of Hegel's Notion and the historical time meant to actualize it. As will become clear, because these fugitive differences are incompatible with identity, their inclusion within a synthetic structure necessarily breaks with idealist dialectics. Deleuze here speaks of a "disjunctive synthesis" that relates differences through their difference or through an enigmatic differenciator acting as their conduit. Linking this discontinuous synthesis to Nietzsche's eternal return, Deleuze holds it to be immanent to any apparently continuous spatial or temporal passage.

A world structured by a time out of joint is also one of simulacra, in which identity and its associated forms of difference are illusory. The dynamics of this decentered system generate illusions or effects of solidity, continuity, and a center. Insofar as immanence demands a synthetic structure that refuses the metaphysical crutch of stable identity, it requires, Deleuze holds, that "all identities are only simulated, produced as an optical 'effect' by the more profound game of difference and repetition" (Deleuze 1994, xix). Simulacra must not be confused with appearances, which have secondary status in relation to essences, and though in some sense they may be "false," this term no longer designates the opposite of truth, since it is no longer part of any "true world." As Nietzsche says, demolishing essence does not turn everything into appearance: "The antithesis 'thing-in-itself' and 'appearance' is untenable; with that, however, the concept 'appearance' also disappears" (Nietzsche 1968, §552). Or, put differently: "There might only be an apparent world, but not *our* apparent world" (§583); "the antithesis of this phenomenal world is not 'the true world,' but the formless unformulable world of the chaos of sensations—*another kind* of phenomenal world, a kind 'unknowable' for us" (§569). For this reason Nietzsche refuses to put stock in scientific empiricism over metaphysical idealism, since the former reaches only the most superficial aspects of existence (see Nietzsche 1974, §373). Instead, he posits a *microscopic* or *virtual* sphere of dynamic relations that generates the fictitious stabilities and seemingly fixed oppositions that both idealism and empiricism seek to ground.[14] These illusions, which are essentially macroscopic emergent phenomena that are irreducible to their microscopic conditions of emergence, are no less real for being simulations, but they do not have the substantiality and durability often attributed to them. Nor do they hold any secondary status, as they arise from the play of simulacra—that is, "systems which relate different to different by means of difference (. . . such systems are themselves simulacra)" (Deleuze 1994, 126). They have "convincing force" (Nietzsche 1968, §488) because the simulacrum, as Lacan says, presents itself not as another appearance but as the truth or essence behind appearances.[15] These simulated stabilities thus give sense to our selves and our world, but they remain inadequate to the ungrounded structure and the dynamics from which they arise. They may prove useful in enabling humans to "endure life. But that does not prove them. Life is no argument. The conditions of life might include error" (Nietzsche 1974, §121). What Nietzsche identifies as the metaphysical errors of reversing cause and consequence and positing false or imaginary causes (Nietzsche 1990, 57–64; also 1968, §229) follow from treating these surface projections as substantial realities, thereby "giving a false reality to a fiction" (1968, §521). Similarly, Deleuze maintains that when treated as more than they are, these simulations generate the transcendental illusions that surround representation, leading thought to transcendence.[16]

14. "If we give up the effective subject, we also give up the object upon which effects are produced. Duration, identity with itself, being are inherent neither in that which is called subject nor in that which is called object: they are complexes of events apparently durable in comparison with other complexes—e.g., through the difference in tempo of the event (rest—motion, firm—loose: opposites that do not exist in themselves and that actually express only variations in degree that from a certain perspective appear to be opposites)" (Nietzsche 1968, §552).

15. "The picture does not compete with appearance, it competes with what Plato designates for us beyond appearance as being the Idea. It is because the picture is the appearance that says it is that which gives the appearance that Plato attacks painting, as if it were an activity competing with his own" (Lacan 1981, 112).

16. "Representation is a site of transcendental illusion" (Deleuze 1994, 265). "The poisoned gift of Platonism," Deleuze states, "is to have introduced transcendence into philosophy, to have given transcendence a plausible philosophical meaning" (1998, 137). Plato makes this possible through his separation of legitimate copies from illegitimate simulacra—a dualism, Deleuze argues, hidden beneath and underpinning the more obvious dualism of Form and copy. Elevating the status of partially stable patterns over the flowing differences that generate them, Plato grants to the former alone an internal resemblance to transcendent Forms, which then ground the stable patterns and enable the dismissal of flowing differences as false pretenders that participate in the Forms only by deception. This

This work's initial ontological concern is time as a discontinuous synthesis that structures change and movement, completes immanence, and generates both a surface of sense and the superficial illusions of identity, stability, and transcendence. But its importance is not limited to ontology. Rather, insofar as time is considered the form of what changes or moves, it directly concerns those domains where seemingly stable identities and apparently continuous transformations refer back to a structure of out-of-sync constitutive relations, disjoined and dispersed through a differenciator and functioning as a surface of sense. On the one hand, there is a dissynchrony that is fundamental to our language, our selves, and the microscopic or virtual relations, forces, and differences immanent to them. For Deleuze, this is the dimension of sense, whose time is that of the eternal return and which exceeds and underpins the structure of propositions that denote objective truths, manifest subjectivities, or signify universal concepts. For Foucault, it is the realm of statements that allow propositions, sentences, and speech acts to "make sense" (Foucault 1989a, 86) and whose modifications do not follow "the temporality of the consciousness" (122). And for Freud and Nietzsche, it is a domain of unconscious drives or instincts that function with no notion of linear time. On the other hand, there are emergent illusions of identity and stability, which organize much of our thinking, our discourse, and our personal, social, and political worlds. There is an experience of unity, of being an ego or an "I," that for Nietzsche accompanies the dynamic of our drives and the becoming dominant of certain drives, but that is easily mistaken for a governing center. There are subject and object positions, upon which knowledge and discourse depend, which for Foucault arise in the intersections of heterogeneous discourses that are interwoven with one another. And there are overall unities of domination, linked to the institutions of law and sovereignty, which Foucault considers the terminal forms of the multiplicity of power relations (Foucault 1990a, 92–93). These apparent stabilities are indispensable to our personal and social lives, but they are not exhaustive. Resistances, Foucault says, are immanent to power relations because they are nothing other than the disconti-

---

sets the condition for the hegemony of representation, which understands difference through the requirements of identity (see Deleuze 1994, 59–64, 264–65, 270; also 1990, 253–66). Consequently, the reversal of Platonism, which Deleuze calls the task of modern philosophy (1994, 59), must reduce identities to the status traditionally given to simulacra, so as to achieve "a restoration of immanence in its full extension and in its purity, which forbids the return of any transcendence" (1998, 137).

Smith (2006) reviews all these points, demonstrating how Deleuze revises the conception of the simulacrum in developing an ontology of the immanent generation of Ideas.

nuities of these relations themselves. There are lines of flight and deterritorializations, for Deleuze and Guattari, that are primary in any formation of desire. The same processes that generate stabilities and identities also serve as the mechanisms by which they are overcome and dissolved.

This overcoming is an ethical and political task. Or, perhaps better, it is an ethical task that flows into politics. It is a crucial task insofar as political and social life continues to privilege fixed markers and identities that are no more than surface projections. On the one hand, as Foucault demonstrates, these markers, in the form of categories of normality and deviancy, coordinate the deployment of modern disciplinary power. Even though the mechanisms and techniques of this power are unable to interpellate or fasten individuals to these positions—and, indeed, Foucault shows that this was never their aim—this does nothing to disrupt the continued extension of disciplinary power and the modern will to truth's demand for certainty. On the other hand, neither Foucault nor Deleuze is particularly interested in a politics that responds to this condition simply by asserting the rights of the marginal, the deviant, or the excluded, as this continues to privilege the idea that establishing collective identity is the precondition for political action. Instead, both Foucault and Deleuze propose ethical and micropolitical projects that seek a different kind of sense, one that identity, opposition, and related terms cannot grasp. The nature of time and its relation to movement and change make this undertaking possible.

These ontological, ethical, and political concerns and goals, which have certainly been all too briefly outlined above, will be pursued through eighteen short reflections, each focusing on a particular philosopher or theme or establishing a conversation between two or three thinkers. They are written to stand independently but also to link with other reflections standing at different proximities to them. Together they constitute a series of interwoven fragments, a structure corresponding to their subject matter. Such disparate pieces may be ordered in many ways, but this work will retain something of a linear character—time, after all, remains on some level linear. Early reflections focus on time, discontinuity, immanence, and sense, addressing these chiefly through engagements with Aristotle, Wittgenstein, Bergson, Bachelard, Deleuze, Hegel, Irigaray, Lacan, and Plato. These are followed by reflections that explore the constitution of a surface that structures discursive formations, linguistic sense, and the psyche, drawing primarily on Foucault, Deleuze, Nietzsche, Freud, Melanie Klein, and ancient Stoicism. The final reflections address nihilism and the will to truth, disci-

plinary politics and power relations, and alternative political and ethical possibilities, engaging with Adorno, Nietzsche, Heidegger, Foucault, and
Deleuze and Guattari.

In all these areas, time's structure is the guarantor of the thinking and
novelty that Nietzsche associates with the revaluation of values. Its discontinuity ensures a break with the past and the overcoming of old and defunct
perspectives. This is, of course, the promise that has led recent political and
ethical philosophy to focus on the nature of time. This work argues that this
promise is fulfilled only at the cost of identity, the ego, and their associated
categories. If identity remains in some way necessary to our understandings
and self-understandings, it is only as a marker useful for coordinating and
organizing certain aspects or levels of life, and only when these are considered in abstraction. It seems to me that identity is still held to be more than
this in many political and philosophical circles aspiring to escape from, or
at least displace and circumscribe, metaphysics, essentialism, and transcendence. I would go so far as to say that the exposure of identity as a simulation is a lesson from Nietzsche, Foucault, and Deleuze that often goes
unappreciated even by theorists who draw inspiration from them.[17] The result, I believe, is that too often identity is held to be historically contingent
and fluctuating, yet still indispensable for politics, ethics, meaning, and
thought. Considerations of the concrete structure of time, however, can
open up another path.

---

17. On this lack of appreciation in contemporary political theory, see Widder (2000;
2002, chapter 1; 2004; and 2005).

1

The "Vulgar" Aristotle

ARISTOTLE IS OFTEN CREDITED with the first rigorous formulation of the "vulgar" conception of time as an infinite series of "nows" stretching from future to past. For many, this ordinary, chronological conception—which reduces time to space,[1] unfolds it in a linear succession,[2] or subordinates it to movement[3]—is a baseline of what contemporary thinking must surpass. Certain difficulties, however, must be negotiated to ascribe this position to Aristotle fully. On the one hand, the discussion of time in *Physics*, book 4, opens by stating that not only are past and future nonexistent, but "the present 'now' is not part of time at all, for a part measures the whole, and the whole must be made up of the parts, but we cannot say that time is made up of 'nows'" (Aristotle 1934–57, 218a). The "now," in this sense, pertains to time, but is not part of time.[4] On the other hand, Aristotle holds that "neither would time be if there were no 'now,' nor would 'now' be if there were no time . . . time owes its continuity to the 'now,' and yet is divided by reference to it" (219b–220a). Here, the "now" is essential to time's being. Taken together, these claims suggest a fundamental metaphysical aporia, whereby time both does and does not exist and the present

1. This is Bergson's primary critique. See Bergson (1910, chapter 2; 1998, 318–19).
2. Heidegger (1982, 242–44, 255) reproaches Bergson for accusing Aristotle of reducing time to space but criticizes Aristotle's conception for ascribing to time a linear and unidirectional succession: "Time as Aristotle expounds it and as it is familiar to ordinary consciousness is a sequence of nows from the not-yet-now to the no-longer-now, a sequence of nows which is not arbitrary but whose intrinsic direction is from the future to the past" (260).
3. Despite his debt to Bergson, Deleuze's criticisms of the ordinary conception of time focus on its treatment of time as the measure of movement, making little reference to time's being spatialized. The transition made from the first to the second cinema book (Deleuze 1986 and 1989), for example, is governed by Deleuze's claim that insofar as time is read off of movement, we are given only an indirect image of what it is.
4. "Thus we have shown that there is a something pertaining to time which is indivisible, and this something is what we mean by the 'present' or 'now'" (Aristotle 1934–57, 234a).

"now" functions ambiguously as both limit and transition, dividing and connecting the sequence of past and future "nows." Aristotle is thus taken to articulate an ordinary conception of time that Hegel's dialectical mediation of time's dimensions through the present "now" completes.[5]

Aristotle, however, seems to give not one account of a "now" with divergent but essential functions, but two very different accounts, one analyzing time in relation to perceived change and another analyzing time itself, without reference to this essential relation.[6] The first gives the "now" a central standing, but not as part of time. In the second, the present "now" constitutes time, but its function conflicts with the nature of indivisibles articulated in connection with the first depiction. Both accounts are necessary for Aristotle's physics, the science of moving things,[7] but rather than articulating the common conception of time, they exceed it while giving it "its rightful due."[8] Already with Aristotle, then, time is disengaged from any subordination to movement, space, and continuity.

The instant corresponds to the spatial point insofar as "time is divided or undivided in the same manner as the line" (Aristotle 1984b, 430b), yet these indivisibles cannot be components of what they divide because they are inconsistent with the nature of a continuum. Aristotle works out the

5. See Hegel (1970, §§257–59). Heidegger holds that with Hegel "the sequence of 'nows' has been formalized in the most extreme sense" (1962, 484) through a "levelling off" that allows the now to "be intuited as something present-at-hand, though present-at-hand only 'ideally'" (483). Moreover, Heidegger maintains that "the priority which Hegel has given to the now' . . . makes it plain that in defining the concept of time he is under the sway of the manner in which time is *ordinarily* understood. . . . It can even be shown that his conception of time has been drawn *directly* from the 'physics' of Aristotle" (500n30). Replying to Heidegger's reading, Derrida (1982) holds that for Aristotle the "now" both is and is not part of time, but he retains the link Heidegger draws between Aristotle and Hegel. Heidegger, of course, also ascribes the ordinary conception of time to Bergson. In defense of Bergson, see Durie (2000).

6. The question is whether this essential relation to change determines what time *essentially* is. There may be an internal connection between time and change, but the definition that seems to follow, holding time to be the number of change, is an abstract and external one and Aristotle has much more to say on the subject. Indeed, Aristotle uses the argument that time requires change only to ground two claims: that time can serve as a measure for change and that both time and change have neither beginning nor end. Neither of these points relates to time's essential structure. On the ontological claims that can be derived from the epistemological premise that time cannot exist without change, see Sorabji (1983, chapter 6), and Coope (2001).

7. Aristotle thus ends book 4 by declaring: "This closes our investigation of time and its properties, *in so far as they are germane to our inquiry*" (1934–57, 224a, emphasis added).

8. Heidegger (1962, 39) announces his own project in *Being and Time* with these words. In his more extensive engagement with Aristotle in *The Basic Problems of Phenomenology* (1982, 232–57, esp. 256–57), he holds that Aristotle goes beyond the common conception, but only by showing how it becomes accessible.

fundamental reasons in two key discussions in *Physics,* that of the infinite or unlimited in book 3 and that of continuity in book 6. Book 3 examines the place of the unlimited, which is clearly a principle of nature (Aristotle 1934–57, 203b), being attributed to natural things (204a). In accordance with teleological constraints, Aristotle insists that the unlimited cannot be associated with extension. There can be no infinite substance or body (204b–205a), no infinitely extended space (205a), no actually infinite magnitude (207b): "the unlimited cannot be a quantum at all" (206a). An infinite magnitude might be conceivable, but it is arguably unnecessary for mathematics[9] and in any event is irrelevant to physics (204a–b). Time extends infinitely forward and backward, but this follows from its relation to movement (207b) and specifically to circular movement (223b), which is complete and, in the case of heavenly rotation, indefinitely repeating (see book 8). Aristotle therefore links the infinite to division: finite quanta of space, time, and motion are all divisible into ever smaller segments, so that the unlimited "exists only . . . as an endless potentiality of approximation by reduction of intervals" (206b).[10] Indeed, divisibility without limit defines the continuum (see 185b, 200b, and 207b). Crucially, however, this unlimited division is a potentiality *that can never become actual:* "with an illimitable potentiality . . . it can never become an unlimited actuality" (206a). Despite certain interpretations, then, finite periods of any continuum, spatial or temporal, are not infinitely divisible but only *potentially* infinitely divisible.[11]

9. "Nor does this account of infinity rob the mathematicians of their study; for all that it denies is the actual existence of anything so great that you can never get to the end of it. And as a matter of fact, mathematicians never ask for or introduce an infinite magnitude; they only claim that the finite line shall be of any length they please; and it is possible to divide any magnitude whatsoever in the same proportion as the greatest magnitude. So that the question under discussion does not affect their demonstrations; whereas actual dimensional existence can only be found in actually existent magnitudes" (Aristotle 1934–57, 207b). Nevertheless, geometrical theorems exist that must postulate infinite extension (see Hintikka 1979, 128–31). Furthermore, the claim that mathematics contains only conceptual but not real magnitudes has little relevance, since Aristotle holds that mathematics deals with physical objects, just not *qua* physical (Aristotle 1934–57, 193b–194a).

10. An unlimited addition of quantities is also conceivable, but it is only "identified by reciprocity with endless division" and "in the sense of exceeding *every* finite magnitude as the result of addition, the unlimited cannot exist even potentially" (Aristotle 1934–57, 206b).

11. Bergson clearly misses this qualification in his early Latin thesis on Aristotle, where he states that for Aristotle "division will go on into infinity" (1970, 70). This erroneous reading continues in later writings where Bergson attributes infinite divisibility to the common conceptions of both time and space, even where he does not attribute these views directly to Aristotle (see Bergson 1983, 142–58; 1991, 206; 1998, 154–57). It is implicit in Heidegger's discussion of the common conception of time, which he traces to Aristotle ("the sequence of 'nows' is uninterrupted and has no gaps. No matter how 'far' we proceed in 'dividing up' the 'now,' it is always now" [1962, 475]), and explicit in various Heidegger and Bergson-inspired readings of Aristotle (see, for example, Durie 2000, 163; and Sadler 1996, 69).

Tnis saves Aristotle from the absurdity of a finite magnitude being composed of infinitely many unextended parts, although it conflicts with a basic
principle of his metaphysics that a potentiality must be able to become actual (see Widder 2002, 62–72).

Book 6 further elaborates the nature of continuity and indivisibles, building on the definition of continuity given in book 5, in which parts are continuous when their extremities not only touch but become one, making the
parts a single thing.

> I mean by one thing being continuous with another that those lim-
> iting extremes of the two things in virtue of which they touch each
> other become one and the same thing, and (as the very name indi-
> cates) are "held together," which can only be if the two limits do
> not remain two but become one and the same. From this definition
> it is evident that continuity is possible in the case of such things as
> can, in virtue of their natural constitution, become one by touching.
> (Aristotle 1934–57, 227a)

The ephemeral nature of these limits indicates that they cannot really be
part of the continuum they divide. However, Aristotle continues in book 6
by maintaining that no continuum can be composed of indivisibles, since
two indivisibles would need to be contiguous or touching, yet one dimen-
sionless point or instant could touch another only by occupying the same
place, and "a continuum is divisible into parts which are distinguishable
from each other in the sense of being in different places" (231b). This reinforces the argument concerning potentially unlimited divisibility, since "if
a succession of indivisibles could make up a continuum either of magnitude
or time, that continuum could be resolved into its indivisible constituents.
But . . . no continuum can be resolved into elements which have no parts"
(231b). Conversely, no extended continuum can be indivisible (223b), making impossible any atomic units of time or space. These considerations
allow Aristotle to refute Zeno's paradoxes by arguing, first, that they presuppose atomic units of time and space (239b) and, later, that they presuppose
an actual infinity of spatial points and temporal instants.[12]

---

12. "If the points are actual it is impossible, but if they are potential it is possible. For one
who moves continuously traverses an illimitable number of points only in an accidental, not
in an unqualified, sense; it is an accidental characteristic of the line that it is an illimitable
number of half-lengths; its essential nature is something different" (Aristotle 1934–57, 263b).
This solution replaces (263a) that proposed in book 6, chapters 2 and 9, which is based on
the argument that as time is infinite in the same sense as movement—that is, infinitely
divisible—an infinite number of points is not really crossed in a merely finite time. Because

What, then, of the two accounts of the "now" in book 4? In the first, the "now" is not part of time's flux. It divides an extended time period, but the division is ephemeral or even only conceptual, like a cut made into flowing water. Time being perceived only with movement or change, but, unlike movement, existing everywhere rather than relating only to a particular moving thing, indicates that "time is neither identical with movement nor capable of being separated from it" (219a). Since movement involves passage from a "here" to a "there," magnitude (i.e., distance) pertains to it, and because magnitude is continuous, so too must be movement and, by extension, time: "for it is because magnitude is continuous that movement is so also, and because movement is continuous so is time" (219a). By analogy with magnitude, movement and time are also quantifiable:[13] "since there is a before-and-after in magnitude, there must also be a before-and-after in movement in analogy with them. But there is also a before-and-after in time, in virtue of the dependence of time upon motion" (219a). Time's continuous magnitude being distinguished by a first and a second "now," corresponding to the first and last limit determining a movement, leads to the conclusion that "this is just what time is, the calculable measure or dimension of motion with respect to before-and-afterness. . . . Time, then, is not movement, but that by which movement can be numerically estimated" (219b). The role of the "now" as a marker of motion—"to mark it [time's flux] is its essential function" (219b)—thereby arises from the way magnitude pertains to time. "It is in virtue of the countableness of its before-and-afters that the 'now' exists; so that the 'now,' wherever found in the before-and-afters, is identical (for it is simply the mark of the before-and-afters in motion), but the before-and-afternesses it marks differ; though the nature of the 'now' depends on the markableness of any before-and-after in general, not on the specific before-and-after marked by it" (219b).

In respect to this counting, Aristotle distinguishes numbering number and numbered number. The number 12, for example, can be treated as

---

the earlier answer expresses no qualms with actually existing infinities, Bostock (1991, 180) argues that book 6 was written before the discussion of infinity in book 3.

13. Aristotle defines magnitude as a quantity that is measurable by division into continuous, numerable segments: "'Quantity' means that which is divisible into constituent parts, each or every one of which is by nature some one individual thing. Thus plurality, if it is numerically calculable, is a kind of quantity; and so is magnitude, if it is measurable. 'Plurality' means that which is potentially divisible into non-continuous parts; and 'magnitude' that which is potentially divisible into continuous parts. Of kinds of magnitude, that which is continuous in one direction is length; in two directions, breadth; in three, depth" (Aristotle 1933–35, 1020a).

either the twelfth mark used to count discrete units or a continuous magnitude divisible into 12 units (or 24 half-units, etc.), these units in no way composing the magnitude itself.[14] In this respect, time is a "concrete numerable" (219b, translator's note a) and "the dimension that is counted" rather than "the counter that counts" (220b), counted "nows" being the numerators of this numerable dimension (220a, translator's note c). Only in being counted by an intellectual consciousness (see 223a) does this numerable become the numerator of diverse movements, just as a straight edge, once marked off, can measure diverse lengths. Something thereby exists in time only insofar as it "exists in it as number (that is to say, as countable)" (221a). This indicates, however, only that the counting of "rows"—or, rather, of the units of time delineated by "nows"—is crucial to the subjective experience of time ("when we experience no changes of consciousness . . . no time seems to have passed" [218b]), while time itself remains grounded in the objectivity of movement, which does not require consciousness (223a). The counting of "nows" is merely the counting of marks. Time, like the straight edge, persists whether or not it is marked.

Aristotle's second account, introduced within the first, treats the "now" as a real, constitutive division. In marking a before and an after, the "now" "differs perpetually, but inasmuch as at every moment it is performing its essential function of dividing the past and future it retains its identity" (219b). Here, the "now" is a kind of moving, living present that continually divides and synthesizes time's dimensions. Aristotle represents it by reference to a moving object, which "may be regarded as a point" and which "retains its identity, but its relations alter. . . . And as time follows the analogy of movement, so does the 'now' of time follow the analogy of the moving object" (219b). In its incessant flux, this present "now" traces a continuum of time, just as "a point . . . constitutes (by its movement) the continuity of the line it traces" (220a). The "now" thereby constitutes time's continuity—"it is through the 'now' that time is continuous" (222a)—but, as with the first account, this continuity is based on an argument by analogy.

Aristotle proposes to link the moving and static "nows" by appealing to the division of potentiality and actuality. The real or actual "now" is in perpetual motion, dividing and uniting, "as the coincident end-term and beginning-term of past and future time" (222a), but potentially it may mark countable segments of time, "and in this potentiality one 'now' differs from

14. See Aristotle (1934–57, 219b); see also Coope (2005, 89–92), Derrida (1982, 58–59), and Heidegger (1982, 239).

another" (222a).[15] This is a "mental potentiality" (222a) corresponding to the consciousness that counts time while perceiving motion. As a stationary and countable limit, the "now" "is not time, but is incidental to time" (220a). But it refers back to the moving "now" dividing and connecting past and future: "these two capacities of potential divider and actual uniter pertain to the same actual 'now' and on the same count of its being two limits at once, but its essential and defined functioning in the one capacity differs from that in the other" (222a). Nevertheless, this attempt to reconcile the two "nows" and their functions is unsustainable within Aristotle's terms, because *an indivisible cannot move or change.* Such change would require the indivisible to pass from one position or state to another—"for every change moves along a definite line from this condition to that" (234b)—but it would thereby have to pass through being partially in one state and partially in the other, becoming divisible by the process (240b). Ironically, such change would be possible, Aristotle acknowledges, only if time were composed of atomic units or indivisibles. Aristotle's "authentic" (234a) or actual present would therefore require that the markers by which time is counted off were not potential or incidental dividers but real constituents of a time that was either composed of atomic units or actually divisible to infinity—a time for which Zeno's paradoxes obtain.[16]

This much follows from Aristotle's analysis: that whatever time is, it can be numbered so as to measure movement, but this counting constitutes only its relation to movement and change for a perceiving consciousness; that in its flux, it can be considered continuous, but only by analogy with magnitude or the moving object tracing a line; and that markability and continuity cannot be resolved to make the former derive from the latter as the incidental derives from the essential. If these abstractions and analogies

15. Richard Sorabji (1983, 46–51) argues that Aristotle never clearly distinguishes static and flowing conceptions of the "now" because he fails to appreciate the difference between the punctuality of instants and the constant character of presentness, and so does not really "ever say that there are two nows, one flowing, one static" (49). Strangely, however, Sorabji does not mention Aristotle's distinction between the "now" considered in actuality and potentiality, except indirectly when he argues that Aristotle does not treat the flowing aspect of the "now" as inessential (50).

16. "For the only hypothesis on which it could be supposed to have any motion while never moving during any period of time would be the hypothesis that time is made up of 'nows,' for then in every 'now' it might be supposed to have moved or to have changed, in such a way as never to be in the process of moving but always in the state of having moved. Now this has already been shown to be impossible; for neither is time made up of 'nows,' nor a line of points, nor motion of 'having-movednesses.' For this is what the assertion of motion being composed of atomic motions amounts to, just as though time could be built up out of 'nows,' or magnitude out of points" (Aristotle 1934–57, 240b–241a).

are removed, however, can Aristotle's thought yield a time that, although it must be experienced as something countable and continuous, is *neither* a sequence of "nows" *nor* a continuum? The former seems already established, but the latter requires probing more deeply into how time is actually rather than potentially divided.

Aristotle insists upon two precise and identifiable limits of time, one relating to processes of change—and so to the marking of a before and an after in time—and the other concerning the structure of time itself. With respect to the first, Aristotle maintains that change terminates at an indivisible instant, but no instant can be identified when change begins. A divisible segment of time during which a change is completed contains a period when the change is still taking place and one when it has already finished. This segment can be divided into smaller segments until reaching a "now" that no longer contains both states. Since a thing cannot be both in the process of change and no longer changing, it must be possible to "speak of the primary 'when' 'at' which the change has been completed (for at that instant it is true to say that the change has been accomplished)" (236a). However, no similar procedure is possible for identifying the beginning of a change, since any instant at which change is said to begin must also be one in which it has begun: "It is clear, then, that there cannot be any irreducible period of time which in its entirety is the 'first' period of the change, since there is no limit to the divisions of a period, and so you can always show that the change was already taking place before the whole of any period, however minute, had passed" (236a). The argument for a definite and identifiable endpoint is based on the example of change between contradictories—where a thing must have finished changing at the moment it changes, since no intermediary position exists between contradictory states—but it is then applied to all forms of change, "for there is no pertinent difference between them" (235b). The argument against any definitive starting point, however, is based on the rejection of instantaneous change (see 234b and 237a), which necessitates that change occurs across an extended continuum of time. While all of this is in tension with Aristotle's understanding of the "proper time" of change being the exact period of its duration (236b), it suffices for his teleology, since "the terminal 'now' is what defines a period, a period being what lies between two 'nows'" (237a).

A specifiable endpoint of a divisible process of change is fully compatible with the time in which it occurs being continuous. The division and unification effected by the actual "now," however, is quite different. In every way, and seemingly despite his overt intentions, Aristotle presents it as a site of discontinuity. Continuity, being defined by potentially infinite divisibility and the melding of limits into unity, requires magnitude and extension. Change cannot be instantaneous because this would entail the paradox of a thing being unchanged and changed at once.[17] In being extended, change is thereby mediated. But the authentic indivisible present establishes an immediate passage. It is a real division of time, rather than a mere marker, "but to make an actual bisection is to effect a motion that is not continuous but interrupted" (263a–b). On the one hand, "the most obvious thing about time is that it strikes us as some kind of 'passing along' and changing" (218b), even though this passing differs from the movement of things in time. On the other hand, the present "now" is "regarded as the limit up to which the past has run, none of the future being this side of it, and also as the limit from which the future runs, none of the past being that side of it" (234a). While nothing can move or be at rest in the present "now," it is undeniable that the passing of future into past does indeed occur there. Such a passing must be immediate, discontinuous, and *ungrounded*. Time owes its continuity to the "now," which constitutes its synthetic structure and allows it to be counted, but it is discontinuous for the same reason.

It might seem, Aristotle notes, that time does not exist, since the past no longer exists, the future does not yet exist, and the present "now" is not a component (217b–218a). Yet time is one of the categories of being (Aristotle 1984a, 1b–2a) and it clearly has a destructive effect on beings existing in it (Aristotle 1934–57, 221a–b, 222b). The true vulgarity of time, its baseness, is perhaps found in this essential destructiveness, which is not in the nature of the other, more pristine categories. Time is excessive in both its infinite forward and backward extension and its foundational discontinuity, making it harmful to any telos. Linking time's infinite extension to the re-entrant circular motion of the heavens addresses only one side of its threat. Any change, teleological or otherwise, that might be smooth, continuous, and mediated, is enframed by the discontinuous synthesis of time. One might therefore say that this discontinuous time structures change as such.

---

17. "For suppose a thing has accomplished the change from A to B at an instant. Then the instant at which it has accomplished the change is not the same as that at which it was in A (otherwise it would be in A and B at the same time)" (Aristotle 1934–57, 237a).

2

Point, Line, Curve

MODERN SET THEORY rejects the *horror infiniti* of earlier thought. An infinite like Aristotle's, linked to endless division of finite magnitudes, is a potentiality that cannot become actual; point and line are incompatible because infinity remains a never completed process of approximation. Conversely, an infinite set, immediately and fully given, offers a way for a continuum to be composed of unlimited indivisibles. The assertion of actual infinities in pure mathematics remains controversial.[1] Nevertheless, its introduction was crucial to the arithmetization of analysis and the grounding of calculus and geometry in number theory.

In *Our Knowledge of the External World,* Bertrand Russell employs the developments of set theory to answer philosophical quandaries concerning spatiotemporal continuity and the infinite. The apparent opposition between discrete points and instants and the continuums they compose results from "a failure to realize imaginatively, as well as abstractly, the nature of continuous series as they appear in mathematics" (Russell 1926, 136). Without necessarily conceding the existence of mathematical points and instants, one can still define mathematical series that are isomorphic with the continuums of time and space: "I do not see any reason to suppose that the points and instants which mathematicians introduce in dealing with space and time are actual physically existing entities, but I do see reason to suppose that the continuity of actual space and time may be more or less analogous to mathematical continuity" (137).[2] This leaves no reason to fol-

---

1. On intuitionist mathematics' rejection of actual infinites, see George and Velleman (2002, chapters 4–5).

2. Also: "although the particles, points, and instants with which physics operates are not themselves given in experience, and are very likely not actually existing things, yet, out of the materials provided in sensation, together with other particulars structurally similar to these materials, it is possible to make logical constructions having the mathematical properties

low Bergson's "heroic methods" (143), which banish mathematics and number from concrete accounts of lived experience.[3] Continuity is conceivable in terms of an infinite series because infinity is not found primarily in endless divisibility. While the ability to divide a finite distance into smaller distances "must be admitted" (141), "it is a mistake to define the infinity of a series as 'impossibility of completion by successive synthesis.' The notion of infinity . . . is primarily a property of *classes,* and only derivatively applicable to series; classes which are infinite are given all at once by the defining property of their members, so that there is no question of 'completion' or of 'successive synthesis'" (160). Set theory defines an infinite class intensionally rather than extensionally, according to its members' properties rather than by enumeration. The inability to reach indivisibles through successive division does not mean that a continuum cannot be composed of them: "But just as an infinite class can be given all at once by its defining concept, though it cannot be reached by successive enumeration, so an infinite set of points can be given all at once as making up a line or area or volume, though they can never be reached by the process of successive division. Thus the infinite divisibility of space gives no ground for denying that space is composed of points" (163).

Given this actual infinity, continuity is defined by the "compactness" of an ordered series. "Continuity, in mathematics, is a property only possible to a *series* of terms, i.e. to terms arranged in an order, so that we can say of any two that one comes *before* the other . . . it does not belong to a set of terms in themselves, but only to a set in a certain order" (Russell 1926, 137–38). A series is "called 'compact' when no two terms are consecutive, but between any two there are others" (138). In the series of rational numbers, for example, "given any two fractions, however near together, there are other fractions greater than the one and smaller than the other, and therefore no two fractions are consecutive" (138). This ensures both the continuity of change and the discreteness of the points and instants through which change occurs. With a true infinity of indivisibles, motion is smooth rather than jerky: "The moving body never jumps from one position to another, but always passes by a gradual transition through an infinite number of intermediaries" (142). Instants can remain discrete because the retention of past moments asserted by Bergsonian duration is explainable

---

which physics assigns to particles, points, and instants. If this can be done, then all the propositions of physics can be translated, by a sort of dictionary, into propositions about the kinds of objects which are given in sensation" (Russell 1926, 147).

3. In reply to Russell, see Durie (2004).

psychologically and physiologically, leaving the mathematical account intact (see 145–47). Zeno's paradoxes depend on conceiving the infinite as either endless divisibility or the endless enumeration of points, a self-contradictory potentiality that can never become actual. Zeno's thesis "assumes the impossibility of definite infinite numbers" (175). But an infinite set resolves these problems. "If *all* the points touched [by a moving body] are concerned, then, though you pass through them continuously, you do not touch them 'one by one.' That is to say, after touching one, there is not another which you touch next: no two points are next [*sic*] each other, but between any two there are always an infinite number of others, which cannot be enumerated one by one" (177).[4]

A difficulty arises, however, regarding the ordering of this infinite series. By using the example of rational numbers, which can be expressed as fractions or ratios of integers, to illustrate compactness, Russell seemingly forgets that mathematical continuity concerns the real number set. The latter includes irrational numbers, which are incommensurable with any integer or fraction, and expresses a higher, nondenumerable power of infinity.[5] Despite holding that "the existence of incommensurables proves that every finite length must contain an infinite number of points" (169) and referring to Cantor's thesis of higher orders of infinity (199), Russell twice answers Zeno's paradoxes with the compactness of fractions (138, 184–85). Irrationals pose two interrelated difficulties for the understanding of continuity. The first concerns the geometric figure from which originates the idea that real numbers must correspond to points on a straight line: the curve descending from a square's diagonal onto a straight line extending from its base.[6] Aside from employing a second dimension to reveal the real numbers

4. Also: "But if, with the mathematicians, we avoid the assumption that motion is also discontinuous, we shall not fall into the philosopher's difficulties. A cinematograph in which there are an infinite number of pictures, and in which there is never a *next* picture because an infinite number come between any two, will perfectly represent a continuous motion" (Russell 1946, 832–33).

5. Cantor's thesis defines the equivalence of two sets by the possibility of putting every member of one into correspondence with one and only one member of the other. This reasoning is used to show that the set of whole numbers {1, 2, 3, . . .} is equinumerous with the subset comprised of every tenth number {10, 20, 30, . . .} (this demonstration that the set of whole numbers has the same cardinality as one of its subsets is the condition of its being an infinite set), and that the set of integers is equinumerous with rational numbers. However, using his "diagonal argument," Cantor demonstrates that no one-to-one correspondence can be established with the set of real numbers, indicating that there are "more" real numbers than rational numbers and that their infinity is of a higher power. The latter is a denumerable infinity, the former is nondenumerable. See Cantor (1955), and Boyer (1991, 566–69).

6. The discovery of irrationals is attributed to the Pythagoreans and the incommensurability of a square's diagonal with its side. If treated as a radius, this diagonal yields a circle

that the straight line is to represent, the figure presupposes the numerical entities it purports to define.[7] The second quandary is that abandoning the geometric aid in favor of a purely arithmetic account seems to leave the place of irrationals in a linear ordered series uncertain, since in their endless, indefinite expansion they are never completely given but only approximated to rational numbers,[8] and in the higher infinity to which they are assigned they exceed any countable or denumerable set.

Before examining Russell's response to these problems, it is necessary to review an earlier outline and answer by Dedekind, whose thesis of cuts seeks to fully determine the place of irrationals in an ordered series and thereby establish the nature of continuity in "a purely arithmetic manner" (Dedekind 1963, 2). While "resort to geometric intuition" is "exceedingly useful" (1), Dedekind argues that it bases irrational numbers "directly upon the conception of extensive magnitudes—which itself is nowhere carefully defined—and explains number as the result of measuring such a magnitude by another of the same kind" (9–10). The lack of correspondence between the rational number set and the points on a straight line "has led to the recognition of the existence of gaps, of a certain incompleteness or discontinuity of the former" (10), compelling an account of completeness able to provide "a scientific basis for the investigation of *all* continuous domains" (10). The incompleteness of rational numbers may be indicated geometrically, but to avoid introducing nonarithmetic notions, "we must endeavor completely to define irrational numbers by means of the rational numbers alone" (10).

---

intersecting the straight line extending from the square's base at a point corresponding to no rational number.

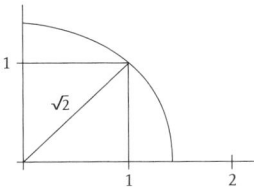

This derivation of irrational numbers from this geometric figure demonstrates that the rational number set cannot account for all the points on a straight line.

7. "In the past, the definition of irrationals was commonly effected by geometrical considerations. This procedure was, however, highly illogical; for if the application of numbers to space is to yield anything but tautologies, the numbers applied must be independently defined; and if none but a geometrical definition were possible, there would be, properly speaking, no such arithmetical entities as the definition pretended to define" (Russell 1937, §264).

8. The place of π, for example, cannot be fully given. Even if calculated to two million decimal places, this merely gives the value of a rational number.

Rather than defining continuity by compactness, Dedekind asserts the singularity of a cut: "If all points of the straight line fall into two classes such that every point of the first class lies to the left of every point of the second class, then there exists one and only one point which produces this division" (11). Every rational number similarly cuts its set into two groups, the first containing all integers and fractions less than the cutting number and the second containing all greater numbers (12–13). The ability of every definitive number to divide its class is considered a basic property of rational numbers (6), but the term "cut" is introduced only when comparison with the points on a line shows the incompleteness of the rational number set (12–13). The incommensurable diagonal descending onto the straight line (9) thereby provides the heuristic device for Dedekind to introduce and extend the notion of cutting to irrational numbers, which are then held also to cleanly divide the real number set (13–19). These irrationals, however, are specified arithmetically, not geometrically, through the proposition of a positive integer D whose square root is not an integer (13). This root divides the number line, so that all rational numbers smaller than the root, when squared, are less than D, while all rational numbers greater than it will be greater than D when squared. The root itself, however, cannot be a rational number, as demonstrated by indirect proof (13–15).[9] These irrational numbers, Dedekind argues, "fill the gaps" among rational numbers and, being fully specified by their cutting of the number series, establish that "the system $\Re$ of all real numbers forms a well-arranged domain of one dimension" (19). Moreover, just as arithmetic operations using rational numbers yield definitive results, the same holds when using real numbers. The sum of two numbers, for example, defined by their cuts, is another, definitive cut, whether they are rational or not (21–24).

Russell maintains, however, that Dedekind's position rests on the "sheer assumption" (Russell 1937, §267) that irrationals are limits that define sections of a continuum. Indeed, while "the mathematical treatment of continuity rests wholly upon the doctrine of limits" (§262), Dedekind's continued use of the geometric image carries the misleading idea that a new number is required for every point corresponding to an incommensurable length on the number line: "From the habit of being influenced by spatial imagination, people have supposed that series *must* have limits in cases where it seems

---

9. The indirect proof first assumes that the square root in question is a rational number that can be put as a ratio of two integers in the form $\sqrt{(a^2/b^2)}$ before demonstrating that the assumption cannot hold for one of the variables.

odd if they do not. Thus, perceiving that there was no *rational* limit to the ratios whose square is less than 2, they allowed themselves to 'postulate' an *irrational* limit, which was to fill the Dedekind gap" (Russell 1971, 71). Nevertheless, "what right have we to assume the existence of such numbers?" (Russell 1937, §267). While irrationals help solve algebraic and geometrical problems, these uses are "powerless to show that [an irrational] is truly a number. They might equally well be regarded as showing the inadequacy of numbers to Algebra and Geometry" (§267). Dedekind's appeal to irrationals as cuts of rational number segments thereby rests on "a logical error" (§267), since a limit cannot be derived from the series it circumscribes; rather, the series must be derived from it. The specification of an irrational suggests a point falling "in between" two series of rational numbers, which the latter approach but never reach, but this "cannot prove that a limit exists, but only that, *if* it existed, it would not be any . . . rational number. Thus irrationals are not proved to exist, but *may* be merely convenient fictions to describe the relations of [two series of rational numbers]" (§267).

Russell therefore rejects the assumption that "the series of real numbers . . . consists of the whole assemblage of rational and irrational numbers, the irrationals being defined as the limits of such series of rationals as have neither a rational nor an infinite limit" (§258). He maintains that "real numbers are really not numbers at all" (§258), but instead are series or segments—essentially equivalence classes of Cauchy sequences—of rational numbers.[10] Some series of rationals, such as {0.9, 0.99, 0.999, . . .} or 1–1/ n (n being some positive integer), converge to a rational limit (see §§260, 319); other segments, corresponding to irrationals, simply have no limit. As the series of segments "has continuity of a higher order than the rationals" and "it can be easily shown that they form a compact series" (§260), segments can be substituted for never properly defined real numbers: "In this doubt as to what real numbers may be, we find that segments of rationals . . . fulfil all the requirements laid down in Cantor's definition, and also those derived from the principle of abstraction. Hence there is no logical ground for distinguishing segments of rationals from real numbers" (§270).

This definition of real numbers explains Russell's reliance on the rational number set to answer Zeno. Continuity does indeed concern the real number set. However, while sufficiently high compactness is ensured by the segments of rational numbers being of a higher infinity and compactness

---

10. Cauchy sequences are sequences whose elements become closer as the sequence progresses.

than the rational numbers themselves, order is established only at the level of rational numbers, which, being denumerable, form "a collection whose terms are all the terms of some progression" (Russell 1937, §278). Through this approach Russell offers an ordinal definition of continuity, designed to avoid reference to space or extension, as a compact nondenumerable series (corresponding to the set of real numbers/rational segments) containing within it an ordered and compact denumerable series (corresponding to the rational number set), where terms of the latter are between any two terms of the former (§277). He then associates this same degree of continuity with spatial continuity (§§416–22). Russell thus accepts that the uncountable infinity associated with real numbers cannot be presumed to be well ordered, but suggests that their order is sustained by a countable infinite subset of rational numbers acting like a skeleton.[11] All limit points of a true continuum are thereby isomorphic with the ordered rational number set, but the continuum also contains an excess of "irrational" terms that do not correspond to any limits and without which the continuum would remain "incomplete" or "imperfect."[12] Perhaps unsurprisingly, Russell continues to refer to the place of irrationals as "gaps" (Russell 1971, 70, 71, 100–101), holding that, because it is "a mistaken belief that there must be 'limits' of series of ratios" (70), one can acknowledge their excessiveness while reconciling

11. Russell's position on continuity was originally developed before Zermelo's formulation of the "multiplicative axiom" or "axiom of choice," which is required for any claim that the set of real numbers is well ordered or that every infinite set has a denumerable subset. Kurt Gödel and Paul Cohen eventually demonstrated that the axiom is independent of the other standard axioms of set theory and therefore is undecidable: both the axiom and its negation are consistent extensions of the Zermelo-Fraenkel axioms, making the axiom of choice neither provable nor disprovable; its adoption or rejection therefore depends on other grounds (see George and Velleman 2002, 84–85). When Russell addresses the issue in later writings, he acknowledges that the axiom has yet to be proved true or false (Russell 1971, 124), but suggests that it can safely be assumed for the lowest infinite set, associated with natural numbers, integers, and rationals: "It may happen that the axiom holds for $\aleph_0$ classes, though not for larger numbers of classes. For this reason it is better, when it is possible, to content ourselves with the more restricted assumption" (129–30). This seems to misunderstand the real import of the axiom and its uncertainty (although Velleman [1993] uses similar reasoning to argue that classical and intuitionistic mathematics can be reconciled at the level of completed infinities such as those associated with rational numbers), but Russell does recognize the consequences if the axiom does not hold: "the continuum or some still more dense series *might* be proved to be incapable of having its terms well ordered" (130).

12. Revising Cantor's definition of a perfect series, Russell holds: "a series is perfect when all its points are limiting-points, and when further, any series being chosen out of our first series, if this new series is of the sort which is usually regarded as defining a limit, then it actually has a limit belonging to our first series" (1937, §274). In other words, with respect to the real number series, all segments have either a rational limit or no limit at all, so that, "in any complete series, either some definable limits do not exist, or the series contains its first derivative [that is, all of its limiting-points]" (§275).

continuity and limits at the lower level of an ordered denumerable infinity. Meanwhile, like Dedekind, he holds that arithmetic operations with real numbers yield definitive results, as though the positions of these "gaps" were clear (see Russell 1937, §269; 1971, 73–74).

Wittgenstein also questions the adequacy of defining irrationals as cuts, maintaining that "the proof of Dedekind's theorem works with a picture which cannot justify *it;* which ought rather to be justified by the theorem" (Wittgenstein 1978, 5.33). This picture is that of the number line extended to real numbers—"the misleading thing about Dedekind's conception is the idea that the real numbers are there spread out in the number line" (5.37).[13] The concept of cutting *"is taken over from the everyday use of language* and that is why it immediately looks as if it had to have a meaning for numbers too" (5.34). While the division of rational numbers is demonstrated without recourse to the picture of an extended line, its indispensability in indicating the continuity of both the line and the set of real numbers shows that it remains impossible to derive irrational numbers without a combination of calculation and geometric construction (5.37). Rather than eliminating the geometric image, Dedekind merely introduces another, wholly fanciful one,[14] resulting in a conception of irrationals that vacillates between the intensional and the extensional.

> The difficulty of looking at the matter now in an intensional, now again in an extensional way, is already there with the concept of a "cut." That every rational number can be called a principle of division of the rational numbers is perfectly clear. Now we discover something else that we can call a principle of division, e.g. what corresponds to $\sqrt{2}$. Then other similar ones—and now we are already quite familiar with the possibility of such divisions, and see them under the aspect of a cut made somewhere along the straight line, *hence extensionally.* (5.34)

Russell's attempt to bypass Dedekind's reliance on the geometrical image still retains assumptions that, for Wittgenstein, are only derived from

13. Also: "The picture of the number line is an absolutely natural one up to a certain point; that is to say so long as it is not used for a general theory of real numbers" (Wittgenstein 1978, 5.32).

14. "The geometrical illustration of Analysis is indeed inessential; not, however, the geometrical application. Originally the geometrical illustrations were *applications of Analysis.* Where they cease to be this they can be wholly misleading. . . . What we have then is the imaginary application. The fanciful application. . . . The idea of a 'cut' is one such dangerous illustration" (Wittgenstein 1978, 5.29).

the image. Russell thereby manages only to ground a picture that is suspect from the start. The residues of Dedekind's erroneous picture persist in two ideas that have no clear sense: that real numbers form a greater infinity than rationals, and that they can, on some level, be ordered.[15] Ultimately, however, the error lies in assuming that a line is composed of points: "Mathematics is ridden through and through with the pernicious idioms of set theory. One example of this is the way people speak of a line as composed of points" (Wittgenstein 1975, §173). This idea produces the "frightfully confusing picture" of a curve that "would have to cut the straight line *between* its points" and so proposes "the concept of a straight line in which a point is missing?!" (Wittgenstein 1978, 5.37). Against this, Wittgenstein maintains that "a line is a law and isn't composed of anything at all" (1975, §173). It is a rule or equation (1978, 5.39) that yields certain results or steps, including results that are not rational.

> A line as a coloured length in visual space can be composed of shorter coloured lengths (but, of course, not of points). And then we are surprised to find, e.g., that "between the everywhere dense rational points" there is still room for the irrationals! What does a construction like that for $\sqrt{2}$ show? Does it show how there is yet room for this point in between all the rational points? It merely shows that the point *yielded* by the construction is *not rational*. (1975, §173)

The irrational too is a rule or law, not a number. Dedekind, asserting that irrationals are "perfectly definite" (1963, 23), merely assumes their status as numbers, but provides a negative account that demonstrates only that they are not rational numbers.[16] But Russell too gives an ultimately negative por-

15. "One pretends to compare the 'set' of real numbers in magnitude with that of cardinal numbers. The difference in kind between the two conceptions is represented, by a skew form of expression, as difference of extension. I believe, and hope, that a future generation will laugh at this hocus pocus" (Wittgenstein 1978, 2.22). Also: "When people say 'The set of all transcendental numbers is greater that [sic] that of algebraic numbers,' that's nonsense. The set is of a different kind. It isn't 'no longer' denumerable, it's simply not denumerable!" (Wittgenstein 1975, §174). Finally: "Surely—if anyone tried day-in day-out 'to put all irrational numbers into a series' we could say: 'Leave it alone; it means nothing'" (Wittgenstein 1978, 2.13; see also 2.16). On Wittgenstein's critique of transfinite number theory's conflation of the higher power of an infinity with its being "greater" in size, see Shanker (1987, 161–75). On how later theories continue Dedekind's confusions, see 187–89.

16. "Indeed, the way the irrationals are introduced in text books always makes it sound as if what is being said is: Look, that isn't a rational number, but still there is a number there. But why then do we still call what is there 'a number'?" (Wittgenstein 1975, §191).

trayal of their nature, holding irrationals to be segments without limits that somehow lie between the rational limits of a denumerable series. Considered positively and intensionally, an irrational is a procedure for generating an ever expanding decimal fraction, "the unlimited technique of expansion of series" (Wittgenstein 1978, 5.19). It is "an arithmetical law which endlessly yields the places of a decimal fraction" (Wittgenstein 1975, §186) and even if it converges toward a definite number, "a *subsequent* proof of convergence cannot justify construing a series as a number" (§197).

> The idea behind $\sqrt{2}$ is this: we look for a rational number which, multiplied by itself, yields 2. There isn't one. But there are those which in this way come close to 2 and there are always some which approach 2 more closely still. There is a procedure permitting me to approach 2 indefinitely closely. This procedure is itself something. And I call it a real number.
>
> It finds expression in the fact that it yields places of a decimal fraction lying ever further to the right. (§183)

At best, "a real number lives in the substratum of the operations out of which it is born" (§185). Yet if one adopts the "good rule that . . . a number is that which can be compared with any rational number taken at random . . . that for which it can be established whether it is greater than, less than, or equal to a rational number" (§191), then an irrational is a number only when approximated to a rational number. On the one hand, an irrational's expansion "runs through a series of rational approximations. When does it leave this series behind? Never. But then, the series also never comes to an end" (§181). The decimal places of an irrational can be extended indefinitely, yet "no matter how far I go with my approximations, there will always also be a corresponding fraction" (§181). On the other hand, even if a set of all irrational numbers save one could be given, the missing number could not be specified, since some other irrational would agree with it through any number of decimal places, however far they were extended.

> Suppose that it's π. If an irrational number is given through the totality of its approximations, then up to *any* point taken at random there is a series coinciding with that of π. Admittedly, for each such series there is a point where they diverge. But this point can lie arbitrarily far "out." So that for any series agreeing with π, I can find one agreeing with it still further. And so if I have the totality

of all irrational numbers except π, and now insert π, I cannot cite a point at which π is now really needed. At *every* point it has a companion agreeing with it from the beginning on. (§181)

Insofar as the comparison with rational numbers does not hold, real numbers cannot correspond to points of a line: "If I don't know how many 9s may follow after 3.1415, it follows that I can't specify an interval smaller than the difference between π and 3.1416; and that implies, in my opinion, that π doesn't correspond to a point on the number line, since, if it does correspond to a point, it must be possible to cite an interval which is smaller than the interval from this point to 3.1416" (§195).

Ultimately, in relation to rational numbers, "the irrational numbers are—so to speak—special cases" (1978, 5.37). "A real number," Wittgenstein suggests, "can be compared with the fiction of an infinite spiral. . . . For my inability to establish on which side it passes by a point, simply means that it is absurd to compare it with a complete (whole) spiral, for with that I would see how it goes past the point" (1975, §192). The irrational is therefore not a point "falling between" densely packed rational number points, but something akin to an unlocalizable vortex. In this respect Wittgenstein may be compared with Deleuze, for whom the irrational, "at the meeting of the curved and straight lines, produces a point-fold" (Deleuze 1993, 18). For both thinkers, the attempt to squeeze irrationals onto the straight line produces a false image of the continuum. "The irrational number implies the descent of a circular arc on the straight line of rational points, and exposes the latter as a false infinity, a simple undefinite that includes an infinity of lacunae; that is why the continuous is a labyrinth that cannot be represented by a straight line. The straight line always has to be intermingled with curved lines" (17).[17] The irrational does not fill in and complete continuity. Instead, it invokes a fundamental discontinuity. It is infinite and uncountable not by virtue of being "more condensed" than any rational infinity, but by going "beyond the limit" of any continuum. It expresses not the actuality of a fully given infinite set but, in Deleuze's terms, a fully real *virtuality*. Deleuze and Guattari refer to this as a domain of "infinite speed," which is infinite precisely by outstripping the limits of

17. In his stance against Dedekind and Russell, Wittgenstein might also be linked to Deleuze in asserting a "minor" mathematics of problematics against the "royal" or "major" mathematics of theoremetization, axiomatization, and arithmetization. Smith (2003) develops this mathematical side of Deleuze's thought and uses it to respond to Alain Badiou's critique of Deleuze, which is inspired by axiomatic set theory.

spatiotemporal continuity. On a plane of immanence that "envelops infinite movements that pass back and forth through it," concepts are the point-folds moving at "infinite speeds" (Deleuze and Guattari 1994, 36). These movements are indeterminate, but not for that reason simply accidental: "Chaos is defined not so much by its disorder as by the infinite speed with which every form taking shape in it vanishes" (118). Concepts are syntheses whose components "are distinct, heterogeneous, and yet not separable" (19). Moreover, contra Russell, these irrational point-folds organize the domain: "Concepts are the archipelago or skeletal frame, a spinal column rather than a skull, whereas the plane is the breath that suffuses the separate parts" (36). In relation to continuous time and space, the concept's intensive movements seem to travel nowhere, but this only demonstrates how they cannot be fully localized in terms of these continuities. Or, put differently, their location is always revealed to be the site of discontinuous excess and a plurality. These characteristics mark concepts as *events*.

Russell also locates constitutive differences in the indivisibles composing his continuum, although his aim is to reconcile real discreteness and continuity. Russell directs his idea against Bergson's claim that real movement is indivisible: "If, for example, I move my hand quickly from left to right, you seem to see the whole movement at once, in spite of the fact that you know it begins at the left and ends at the right. It is this kind of consideration . . . which leads Bergson and many others to regard a movement as really one indivisible whole" (Russell 1926, 145). The limits of sense experience, however, suggest that "there must be among sense-data differences so slight as to be imperceptible" (150), meaning that such data may be "composed of mutually external units" (152). In short, experience "does not prevent us from admitting that sense-data have parts which are not sense-data" (156), that there is a constitutive realm discontinuous with experience. Similarly, Wittgenstein holds that "continuity in our visual field consists in our not seeing discontinuity" (1975, §137). A strip of alternating black and white patches, for example, appears a continuous grey when the patches are so thin as to be indistinguishable (§137). But this means that continuity is experienced when "we just see *no* parts, *no* leaps (not infinitely many)" (§144). The issue here is how to understand this microscopic, virtual, and constitutive realm beneath apparent continuity: as Russell's exclusive points and instants or as the irrational anomalies, spirals, and folds of Wittgenstein, Deleuze, and Guattari? The question concerns the type of discontinuity constituting and thereby enframing a continuum.

3

Immanence and Sense

IN HIS 1954 REVIEW of Jean Hyppolite's *Logic and Existence*,[1] Deleuze sets the direction for his subsequent work in relation to Hegelian dialectics. Affirming both Hyppolite's reading of Hegel and his criticisms of anthropological readings like Kojève's (1969), Deleuze praises Hegel for demanding that philosophy be an ontology of sense: "*Philosophy must be ontology, it cannot be anything else; but there is no ontology of essence, there is only an ontology of sense*" (Deleuze 1997b, 191). With this gesture, Deleuze places his thought and Hegel's on a common terrain with respect to a single question: what concept of difference is needed for an ontology of sense to be adequate to a philosophy of immanence?

Hyppolite maintains that Hegel's concept of speculative contradiction eliminates all reference to a second world, so that "immanence is complete" (Hyppolite 1997, 176).[2] He further criticizes Bergson, Spinoza, Leibniz, and Hume—who are all marshaled by Deleuze against Hegel—for relying on conceptions of difference that, being less than contradiction, are inadequate to the demands of immanence. Deleuze, however, reverses this criticism: although Hegel shows that philosophy must be an ontology of sense, his negative notion of contradiction is insufficient to the task. It is dialectical contradiction that is less than difference and, failing on its own terms to provide a concrete unity of being and expression, completes immanence only in abstraction. This failure shows how Hegelian sense, for Deleuze, rests upon an abstract conception of difference. The concomitant decon-

1. The importance of Hyppolite's reading of Hegel for Derrida, Deleuze, and Foucault has been noted by Leonard Lawlor, who has detailed this importance with respect to Derrida. See Hyppolite (translator's preface, ix–xv) and Lawlor (2002, 88–104).

2. See also Lawlor (2002, 89): "Hegel's philosophy, for Hyppolite, completes immanence without eliminating difference."

struction of Hegel's thought at the point Hyppolite highlights—the transition from the temporal and historical dialectic of the phenomenology to the eternal becoming of the logic—clears the path for a positive, nondialectical difference able to complete immanence and underpin a concrete ontology of sense. This leads Deleuze to pose the question: "can we not construct an ontology of difference which would not have to go up to contradiction, because contradiction would be less than difference and not more?" (Deleuze 1997b, 195). In this respect, Deleuze can be seen to both rival and complete Hegel's project of immanence—just as Deleuze sees Nietzsche as rivaling and completing Kant's project of critical philosophy (Deleuze 1983, 1, 87–97)—appreciating, at the same time, that Deleuze's work may be seen to rival and complete any number of philosophical projects.[3]

What is meant by "sense"? Hyppolite quotes Hegel to explain the term's ambiguity: "Sense is this wonderful word which is used in two opposite meanings. On the one hand it means the organ of immediate apprehension [i.e., the sense of smell], but on the other hand we mean by it the sense, the significance, the thought, the universal underlying the thing. And so sense is connected on the one hand with the immediate external aspect of existence, and on the other hand with its inner essence" (Hegel, quoted in Hyppolite 1997, 24). Sense thus invokes divergent realms of thought and thing, concept and object, universal and singular. It also, however, refers to *surface*. A surface covers or hides what is underneath, even if there is really *nothing* underneath to hide. Here, sense at least initially opposes essence or being, although it may eventually sublate being or show itself to be the essence of being, leaving nothing but a surface hiding only that there is

---

3. See, for example, Smith's (1997) excellent study of Deleuze's philosophy as a rival to and completion of Kant's. Some of Deleuze's best-known remarks suggest that a positive relation to Hegel is impossible. He not only confesses personal abhorrence for Hegel—"what I most detested was Hegelianism and dialectics" (1995, 6)—but also declares at the end of his famous analysis of Nietzsche: "There is no possible compromise between Hegel and Nietzsche. Nietzsche's philosophy . . . forms an absolute anti-dialectics and sets out to expose all the mystifications that find a final refuge in the dialectic" (1983, 195). Commentators examining the relationship between Hegel and Nietzsche often see Deleuze's *Nietzsche and Philosophy* (1983) as a prominent reading of Hegel whose inaccuracies must be critically exposed and rebutted. Two studies that approach Hegel and Nietzsche very differently but similarly attack Deleuze's reading of Hegel in relation to Nietzsche are Houlgate (1986) and Jurist (2000). Commentators also frequently accuse Deleuze's work more generally of failing to understand the nature of Hegelian dialectics, resulting in his own thinking falling back into the oppositional thought it seeks to surmount. On this point see Descombes (1994) and Malabou (1996). Hardt (1993) holds that Deleuze's initial attacks on Hegel are crude and oppositional but that his later thought "matures" and adopts a strategy of more indirect attack.

nothing to hide: "There is nothing to see behind the curtain . . . the secret is that there is no secret" (Deleuze 1997b, 193). But "surface" also refers to the plane that relates heterogeneous domains, as the ocean's surface divides but also connects water and air. No matter how thin it is, the surface remains different from these realms, even while they are nothing without the surface that delimits them.[4] Surface sense is here posed as an *excess*. Hegelian sense brings all these diverse aspects into play, together with the additional meaning—found in the French *sens* and the German *Sinn*, but not the English term—of direction:[5] sense is the surface that divides, holds together, and *constitutes* by synthesizing, leaving nothing outside of sense in either the depths of things or the heights of Ideas. Even nonsense is therefore part of sense. Moreover, Hegelian sense is *expressive:* sense expresses being, even while no being exists prior to or outside of this expression. Indeed, sense is the self-expression of being: it is a language of being that differs from, even though it may appear within and speak through, empirical or human language.

An ontology of sense overcomes the remaining aporias of metaphysical thought: between the empirical and the essential, the singular and the universal, subjective certainty and objective truth, the subject who speaks and the object about which one speaks. It surmounts both forms of the "ineffable," the absolutely individual and the universal, which seem to escape linguistic expression and which philosophies of transcendence have always taken as limits to knowledge (see Hyppolite 1997, part 1, chapter 1). It thus reaches "actuality . . . the concrete unity of essence and appearance, the presentation which presents only itself and tests its necessity not in a separate intelligibility, but in its own movement and development" (4). Sense is therefore not anthropological: "Anthropology wants to be a discourse *on* man. It assumes, as such, the empirical discourse *of* man, in which the one who speaks and that of which one speaks are separated. Reflection is on one side and being on the other" (Deleuze 1997b, 192). Whatever lacks the concreteness of actuality is abstract and one-sided (Hyppolite 1997, 5). Rather than grasping the totality of what is, empirical and anthropological discourses take part of reality and make it the whole. An ontology of sense moves beyond the empirical, toward the many mutually imbricated layers

4. "Sense is never only one of the two terms of the duality . . . it is also the frontier, the cutting edge, or the articulation of the difference between the two terms, since it has at its disposal an impenetrability which is its own and within which it is reflected" (Deleuze 1990, 28).

5. See Hyppolite (1997, "Translators' Note," xvii–xviii).

of reality residing within the empirical but going unnoticed by empirical thinking—layers that, while different, are not exterior to one another but rather folded into one another. This is why sense, although exceeding both the empirical and the conceptual, does not form a second world transcending our own but instead grasps the totality of all there is, even while grasping itself as something more than an empirical or conceptual totality. Sense is immanent to our world, but it resides within it as something different from the world's immediate appearance. For this reason sense must present itself in the internal passage from one side of the divide to the other, in the movement from the empirical to the conceptual and back.

Hyppolite's concern is where to locate the final sublation of the empirical into the Absolute and, consequently, how to understand the reverse movement by which the Absolute actualizes itself concretely. The *Phenomenology* offers two possibilities. The first is the transition to self-consciousness, which completes the dialectic of consciousness. Here, upon reaching the stage of "Understanding," consciousness remains detached from its world, grasping reality through laws that never fully reconcile universal and singular. The means to surpass this aporia, however, are found in understanding itself, which, by introducing the concept of "Force," knows the unity of an object to be the product of its negative relations to others. Going beyond the object as a thing prior to its relations, consciousness recognizes that, in being separated from its object, it is in fact negatively related to it and so always already part of its identity. The subject thereby finding itself in its object, the promise of Hegel's Introduction is fulfilled: that when consciousness is no longer burdened by some alien "other," "it will signify the nature of absolute knowledge itself" (Hegel 1977, §89). The remaining chapters of the *Phenomenology* then detail the actualization of the Absolute in abstract (the dialectics of self-consciousness and reason) and eventually concrete (the dialectics of spirit, religion, and absolute knowing) accounts of human history.

Hyppolite, however, argues that this reading illicitly makes human self-consciousness the Absolute Subject, resulting in an anthropological account that Hegel rejects: "In the *Phenomenology*, Hegel does not say man, but self-consciousness. The modern interpreters who have immediately translated this term by man have somewhat falsified Hegel's thought. Hegel is still too Spinozistic for us to be able to speak of a pure humanism; a pure humanism culminates only in skeptical irony and platitude" (Hyppolite 1997, 20). Phenomenology and human history present a being that speaks or expresses the sense of itself, but it remains a merely human being, not the being of the

Absolute. Hyppolite therefore turns to a second appearance of the Absolute: at the transition to the *Logic* that closes the *Phenomenology*. Under this reading, a gap remains between subject and object, between what is said and the sense of what is said, throughout the accounts of human experience and history. Although it elaborates the self-reflexivity of the Absolute, "the *Phenomenology* studies the anthropological conditions of this reflection; it starts from human, properly subjective, reflection in order to sublate it, in order to show that this *Phenomenology*, this human itinerary, leads to absolute knowledge, to an ontological reflection which the *Phenomenology* presupposes" (34). Human experience and history are therefore merely the focal points where a self-determining Absolute beyond man realizes itself in time and space. While the anthropological readings remain restricted to the *Phenomenology*, Hyppolite's account brings together all aspects of Hegelian thought: logic, phenomenology, and the philosophy of nature. It also secures the special place of philosophical language. The discourses of mathematics, understanding, and poetry use linguistic signs to signify empty concepts, merely empirical concepts, or to make a show of escaping conceptuality altogether. Philosophical logos, however, presents a language that, in expressing its own sense, concretely expresses the unity of thought and being, of universal and singular, thereby surpassing human thought and grasping the self-expression of the Absolute (see part 1, chapter 3).

But here a new problem emerges. For unlike the passage from consciousness to self-consciousness, where the means for sublation are contained in consciousness's understanding, if the same were true of the passage from history to the Absolute, it would affirm the anthropological thesis that the Absolute is simply the realization of human self-reflection and self-knowledge at the culmination of history. At stake is the question of why Hegelian logos is not another metaphysics: if the Absolute is attained only by transcending the human, phenomenological, and empirical, how can one avoid the conclusion that the logic is another essence, a truth behind appearances? Hyppolite answers that the Absolute is neither substance nor essence but difference and mediation, which maintains dualisms such as subject/object without being held to them (1997, 60–61). As such, logos, phenomenology, and nature all effectively mediate one another, though in different ways: nature presents spirit and logos immediately and without reflection; spirit connects logos and nature through reflection, but the reflection remains finite; and logos is the infinite unity of immediacy and reflection, making it the highest form of the Absolute (103–4). Nonetheless, "it is true that the historicity of this absolute knowledge poses at the very heart of Hegelianism new and perhaps unsolvable problems" (36). Only by virtue of the Absolute—and therefore the logos beyond history—can history have meaning and direction. Historical forms of self-consciousness develop toward reconciliation and reciprocal recognition, so that history, "in temporal dispersion, incarnates this supreme sublation that is the absolute Idea" (185–86). But the logos and its self-determining movement are eternal rather than historical. And while the historical can still be conceived as the self-negation of the eternal, there is no negation internal to history that reunites it with the eternal: "History is indeed the place of this passage, but this passage is not itself a *historical fact*" (189). Nonetheless, it is claimed, the sense of this eternal Absolute "is not another world behind history" (188).

Here, for Deleuze, the possibility for an alternative ontology of sense arises. On the one hand, "the relation between ontology and empirical man is perfectly determined, but not the relation between ontology and historical man" (Deleuze 1997b, 194); on the other hand, the lack of an internal passage from history to logic "assumes, at the least, not only that the moments of the Phenomenology and the moments of the Logic are not moments in the same sense, but also that there are two ways of self-contradiction, phenomenological and logical" (195). The persistent gap between phenomenology and logic signals the failure of Hegel's philosophy of sense, as it reinstates transcendence through its equivocation between the sense of the logic and the sense of history: at best history can have a sense only analogous to the logic, yet it is supposed to incarnate the logic. Hegel's Absolute thus remains an abstraction, which is further highlighted by the distinction Hyppolite draws between the concrete and the actual, a distinction that, absent any dialectical mediation between them, seems only to invoke a standard metaphysical idealism: "Logic is not concrete truth, that of the Idea in nature or in spirit, but the pure truth, the development of the concept in its actuality and of actuality in its concept, the life of the concept" (Hyppolite 1997, 164).

Deleuze therefore locates this failure in Hegel's logic, which understands the being of difference in terms of contradiction: "Being can be identical to difference only insofar as difference is carried up to the absolute, that is, up to contradiction. Speculative difference is the Being which contradicts itself" (Deleuze 1997b, 195). The subsequent question is whether contradiction is an adequate expression of being: "is it the same thing to say that Being expresses itself and that it contradicts itself?" (195). Is there, in short, a difference adequate to the requirements of an ontology of sense that differs from Hegelian contradiction?

4

A Discontinuous Bergsonism

FROM ARISTOTLE TO BERGSON, time's continuity is consistently derived through analogy with local motion: the simple, undivided movement of an object tracing a line; the lifting of a hand.[1] Conversely, the metaphor of music—with its varied layers of rhythm and tempo, melody and counterpoint, staccato and legato—renders fullness the effect of "hatched lines of discontinuity" (Bachelard 2000, 122). Bergson certainly employs musical imagery, comparing the experience of duration to recalling "the notes of a tune, melting, so to speak, into one another" (Bergson 1910, 100), and he calls attention to the mass of continuous sound heard when listening to a foreign language being spoken (1991, 109–16, 118, 121–23). Yet clearly he never moves beyond simple melodies and the stream of foreign words sounds continuous only to untrained ears. Despite his protestations against reducing time to space, Bergson clearly prefers the visual and tactile over the other senses, because the former give uninterrupted fullness to experience.[2]

1. The latter is Bergson's frequent example (see Bergson 1956, 114, 259–60; 1983, 142–44; 1991, 188–91; and 1998, 90–91, 299–300). Aristotle holds that "movement is clearly one of the things we think of as 'continuous'" (Aristotle 1934–57, 200b), maintaining that movement's unity is constituted by "its uninterrupted continuity" (228a). Moreover, "local movement . . . takes natural precedence of the others" (243a; see also 208a–b) and, in comparison with quantitative or qualitative changes, it "is the one which takes its subject least away from its essential nature" (261a).

2. Russell (1946, 826) states: "Bergson is a strong visualizer, whose thought is always conducted by means of visual images." Lawlor (2003, 5–6) and Mullarkey (2005, 482) comment similarly. Bergson (1991, 196–97) writes: "A body, that is, an independent material object, presents itself at first to us as a system of qualities in which resistance and color—the data of sight and touch—occupy the center, all the rest being, as it were, suspended from them. Yet the data of sight and touch are those which most obviously have extension in space, and the essential character of space is continuity. There are intervals of silence between sounds, for the sense of hearing is not always occupied, between odors, between tastes, there are gaps, as though the senses of smell and taste only functioned accidentally: as soon as we open our eyes, on the contrary, the whole field of vision takes on color; and, since solids are

Bachelard's *The Dialectic of Duration* sets out to conceive a discontinuous Bergsonism: "of Bergsonism we accept everything but continuity. . . . We wish therefore to develop a discontinuous Bergsonism, showing the need to arithmetise Bergsonian duration" (2000, 28–29). This undertaking is hardly straightforward, given Bergson's affirmation of difference and heterogeneity. Yet while Bergson consistently defines duration as a continuous succession of irreducibly different qualitative states, Bachelard maintains that "Bergson often tones down this heterogeneity so that as a result, succession seems like a change where things fade and merge into one another . . . the addition of the idea of continuity to that of succession is gratuitous and without proof" (42–43). Moreover, although Bergson occasionally acknowledges that his examples of simple movement really involve imperceptible complexes of contractions, tensions, and forces working with and against each other to create the appearance of undivided motion, he never allows such real, discontinuous complexities to disrupt his thesis of fundamental continuity.[3] Once appended to heterogeneity, continuity is sustained by "a

---

necessarily in contact with each other, our touch must follow the surface or the edges of objects without ever encountering a true interruption." See also Bergson (1998, 168), where instinct is opposed to intellect and related to vision and knowledge at a distance; Bergson (1983, 139–42), where metaphysics and intuition are both linked to vision, traditional metaphysics being criticized for misunderstanding the movement and change grasped by the senses; and Bergson (1999, 31), where melody is considered too complex and fractured to be an appropriate image for duration. Perception of bodies and colors are the two examples of continuous manifoldness given by Riemann (1873, 1.2), which perhaps supports Deleuze's hypothesis that Bergson drew his general problematic from Riemann's distinction between discrete and continuous multiplicities (Deleuze 1991, 39–40, 79). Nevertheless, one must remember that Riemann's continuous manifolds concern space and measure and that Riemann remains agnostic about whether continuity applies at the microscopic level: "it seems that the empirical notions on which the metrical determinations of space are founded . . . cease to be valid for the infinitely small. We are therefore quite at liberty to suppose that the metric relations of space in the infinitely small do not conform to the hypotheses of geometry" (1873, 3.3). On Bergson's relation to Riemann, see Durie (2004). Ansell Pearson (2002, 12) holds that Bergson privileges hearing and that sight and touch do not show us continuity but instead set the conditions for action by dividing up the continuity of the world. Bergson, however, maintains that our orientation toward action does not build on our visual and tactile experiences but imposes external conditions on them and takes them off course: "Our needs are . . . so many searchlights which, directed upon the continuity of sensible qualities, single out in it distinct bodies. They cannot satisfy themselves except upon the condition that they carve out, within this continuity, a body which is to be their own and then delimit other bodies with which the first can enter into relation, as if with persons" (1991, 198). This privileging of the visual and tactile seems to confirm Bachelard's claim that "continuous Bergsonism will move imperceptibly and inevitably to an unforeseen consequence: matter is said to fill time even more surely than it does space. Surreptitiously, the phrase *to have duration in time* has been replaced by the phrase *to remain in space*, and it is the crude intuition of fullness that gives the vague impression of plenitude" (2000, 45).

3. "Let us consider a very simple act, like that of lifting the arm. Where should we be if

whole family, an entire closed cycle of metaphors that will constitute the
language of continuity, the song and indeed the lullaby of continuity" (121).
Although Bergson attributes creativity to time, Bachelard holds that "the
creative value of becoming is limited by the very fact of fundamental continuity" (24). Creativity coming from the impulse of the *élan vital*, of the
virtual past compressed and propelled forward, allows no independent role
for the present instant: "the present can do nothing . . . [it] can create nothing. It cannot add being to being" (24–25). All this, Bachelard contends,
demonstrates the necessity of a negative difference immanent to the heterogeneities of continuous duration: "a void must be postulated between the
successive states characterising the psyche's development, even if this void
may be simply a synonym for the difference between states that are differentiated" (91). Two key moments comprise Bachelard's critique. First, against
Bergson's dismissal of the instant as part of a false, spatialized conception
of time, Bachelard demands priority for the instant understood as "a pure
event" (58). Second, against duration's visible continuity, Bachelard opposes
a microscopic or quantum domain of divergences, discontinuities, and vibrations that Bergson's simple movement-image effaces: "When I say that
a phenomenon taken as a whole *changes* from state A to state B, what I
mean is that between A and B there are myriad details and accidents which
I ignore but which it is always in my power to indicate" (76).[4]

Echoing Aristotle, Bergson maintains that the present instant, taken as a
mathematical point, "is a pure abstraction, an aspect of the mind. . . . You
could never create time out of such instants any more than you could make
a line out of mathematical points" (Bergson 1983, 151). And again: "every

---

we had to imagine beforehand all the elementary contractions and tensions this act involves,
or even to perceive them, one by one, as they are accomplished?" (Bergson 1998, 299).

4. Ansell Pearson argues against Bachelard that Bergsonian duration already contains
discontinuity, privileging radical contingency and the "incommensurability between what
goes before and what follows" (Bergson 1998, 29, quoted in Ansell Pearson 2002, 74), and
that it develops creatively in terms of "the dissociation of tendencies and the divergency of
lines of evolution" (ibid., 88). This reply, however, underplays the thrust of Bachelard's critique, which introduces microscopic discontinuities, not the extended discontinuities Bergson recognizes as abstractions (Grosz [2004, 279n9] also misses this point when defending
Bergson). It also overlooks Bachelard's rejoinder to divergence in evolution: "No doubt there
are halts and there are failures; for Bergson though, the cause of failure is always external. It
is matter that opposes life, colliding with life's momentum which it slows or deflects in a
downward curve" (Bachelard 2000, 40). Moreover, Ansell Pearson's association of Bergson
with Deleuze and of Bachelard with Badiou (2002, 70–72) ignores Deleuze's own break with
Bergson on the issue of virtual continuity. Al-Saji (2004) argues that Bergson's duration
contains discontinuity, but does not address Bachelard. Aitken (2004) suggests a rapproche-
ment between Bachelard and Bergson but uncritically aligns Bergson and Deleuze.

duration is thick; real time has no instants" (Bergson 1999, 36). But Bachelard's instant is not a marker of a continuum. It is an immanent as against a transitive time (Bachelard 2000, 105) and a time of thought and choice running perpendicular to durational time. The time of thought is never synchronous with life (36, 38), yet it gives living duration its consistency: "the cohesion of our duration is made from the coherence of our choices" (37). Vital impulse cannot drive duration, because "what has most duration is what is best at starting itself up all over again" (20) and this requires a founding decision that is "not homogenous with what follows it" (58). If the Cartesian cogito refers to a phenomenal being existing in transitive time, it also implies a second-order cogito, an *I think that I think*, which is affirmed prior to the *I think, therefore I am* (108–9). But this in turn presupposes a third order of the cogito as form (110–11), revealing a becoming within the instant: "this formal becoming rises above and overhangs the present instant; it is latent in every instant that we live" (110). The form of the cogito is synthetic, combining diverse layers of the self living divergent and discontinuous times, continuity resulting from temporal superimpositions. This is no mere contraction of levels of duration, because the instant's becoming entails negativity or interval within it (117). Other psychological states may be similarly described as assemblages of diversity (112–20), including those deep-seated psychic states that Bergson holds, in *Time and Free Will*, most fully illustrate duration's unquantifiable character (Bergson 1910, 7–20).[5] Bachelard's instant provides a depth to time different from duration's thickness: "time has several dimensions; it has density. Time seems continuous only in a certain density, thanks to the superimposition of several independent times" (Bachelard 2000, 102). The instant contains a dialectic of duration, which is not logical but temporal (42) and which is littered with lacunae populating both lived time and the higher orders of thought time (103–4, 112): "Here, we are breaking up Bergsonian continuity, preferring a hierarchy of instants" (37).

Bachelard's instant, however, is only the first discontinuity he introduces into time. The diverse durations the instant brings together are themselves fractured. Durations must be "translated into the language of frequencies" (Bachelard 2000, 78). While their macroscopic appearance may be continu-

---

5. It is noteworthy that Bergson acknowledges that such deep-seated psychic states might not even exist: "But certain states of the soul seem to us, rightly or wrongly, to be self-sufficient, such as deep joy or sorrow, a reflective passion or an aesthetic emotion. Pure intensity ought to be more easily definable in these simple cases, where no extensive element seems to be involved" (1910, 7–8).

ous, on the microscopic level they are nothing but vibrations: "The thread of time has knots all along it. And the easy continuity of trajectories has been totally ruined by microphysics" (81). Bergson speaks similarly when, after *Time and Free Will*, he proposes a reconciliation of quantity and quality by maintaining that the continuous flow of qualitative sensations in consciousness and the extended movements occurring in divisible space both involve the contraction of innumerable vibrations (see Bergson 1991, 202–8; 1998, 300–301). The thesis certainly accords with Bergson's idea, supported by Bachelard, that no mobile lies beneath movement (Bergson 1983, 147). The question, however, is whether Bergson's definition of continuity as the complete interpenetration of differences can include within it the notion of constitutive vibrations. Bachelard insists it cannot. Vibration is inseparable from alternation, negation, and hence discontinuity: "We must ascribe fundamental duality to time since the duality inherent in vibration is its operative attribute" (Bachelard 2000, 138). The quantum nature of these vibrations, however, also subverts the central tenet of duration that the past is the impulse driving time's creativity: "in microphysics, antecedent duration does not *propel* the present and . . . the past does not weigh upon the future" (76). This reinforces the independent power of the present, which not only contracts the past but replaces successive efficient causality with immediate formal causality (76). It also introduces number and quantity into duration: "Qualitative becoming is very naturally a quantum becoming" (102). This quantity is not cardinal number but "ordinal probability," a quantitative difference "marked by the simple interaction of plus and minus signs" (99).

Now, particularly in *Matter and Memory*, Bergson advances his own thesis of multiple levels of duration, just as he advances one concerning constitutive vibrations. Conversely, in *Duration and Simultaneity*, he dismisses both notions. In each case, however, Bergson's maneuvers are governed by the primacy he accords quality over quantity in affirming unquantifiable duration as the ground for countable time: "Ageing and duration belong to the order of quality. No work of analysis can resolve them into pure quantity" (Bergson 1999, 124); also, "it is through the quality of quantity that we form the idea of quantity without quality" (Bergson 1910, 123); and finally, "it must not be forgotten that quantity is always nascent quality: it is, one might say, its limiting case" (Bergson 1983, 191). Thus, in *Matter and Memory*, although microscopic vibrations demonstrate that there are, "beneath the apparent heterogeneity of sensible qualities, homogeneous elements which lend themselves to calculation" (1991, 204–5), these elements are uncountable *except* by a duration sufficiently more relaxed than that given to human consciousness as to be able to tally them.[6] This consideration leads directly to Bergson's positing multiple rhythms of duration (206–8), thus linking the two theses together. Multiple durations are required to ensure that quality retains priority over quantity, once vibrations with frequencies outrunning human perception are admitted.

In *Duration and Simultaneity*, however, Bergson dismisses vibrations as a conventional construct of physicists[7] before proceeding to hold, against relativity physics, the absolute place of the conscious individual.[8] This forms a prelude to Bergson's argument against Einstein: that with respect to human consciousness at least,[9] there can be only one real time, one pace of lived duration. This is based on "an argument by analogy that we must regard as conclusive as long as we are offered nothing more satisfactory . . .

6. "The duration lived by our consciousness is a duration with its own determined rhythm, a duration very different from the time of the physicist, which can store up, in a given interval, as great a number of phenomena as we please. In the space of a second, red light . . . accomplishes 400 billion successive vibrations. If we would form some idea of this number, we should have to separate the vibrations sufficiently to allow our consciousness to count them or at least to record explicitly their succession. . . . A very simple calculation shows that more than 25,000 years would elapse before the conclusion of the operation. Thus the sensation of red light, experienced by us in the course of a second, corresponds in itself to a succession of phenomena which, separately distinguished in our duration with the greatest possible economy of time, would occupy more than 250 centuries of our history. Is this conceivable? We must distinguish here between our own duration and time in general" (Bergson 1991, 205–6).

7. "Colors would probably appear differently to us if our eye and our consciousness were differently formed; nonetheless there would always be something unshakably real which physics would continue to resolve into elementary vibrations. In brief, as long as we speak only of a qualified and qualitatively modified continuity, such as colored and color-changing extension, we immediately express what we perceive, without interposed human convention—we have no reason to suppose that we are not here in the presence of reality itself. . . . [Physics] dissolves the body into a virtually infinite number of elementary corpuscles; and, at the same time, it shows us this body linked to other bodies by thousands of reciprocal actions and reactions. It thus introduces so much discontinuity into it, and, on the other hand, establishes between it and the rest of things so much continuity that we can gather what there must be of the artificial and conventional in our division of matter into bodies" (Bergson 1999, 25).

8. "There is only one motion . . . which is perceived from within, and of which we are aware as an event in itself: the motion that our effort brings to our attention" (Bergson 1999, 26). Also: "Once the ether has vanished along with the privileged system and fixed points, only relative motions of objects with respect to one another are left; but, as we cannot move with respect to ourselves, immobility will be, by definition, the state of the observatory in which we shall mentally take our place" (27).

9. Consciousness need not be anthropomorphic: "We may perhaps feel averse to the use of the word 'consciousness' if an anthropomorphic sense is attached to it. But to imagine a thing that endures, there is no need to take one's own memory and transport it, even attenuated, into the interior of the thing" (Bergson 1999, 33).

[that] all human consciousnesses are of like nature, perceive in the same way, keep in step, as it were, and live the same duration" (Bergson 1999, 32). Peter on earth and Paul in his spaceship must each live the same kind of real time, as "a time lived and recorded by a consciousness is real by definition" (51). Each observer, taking his own frame of reference as absolute, attributes to the other a slower duration, but this is "conventional" (62), a "mathematical fiction" (20), an imaginary time imposed by each one viewing the other's system externally. The result, Bergson argues, is that given the reciprocity of movement in special relativity, one observer will not age two years while the other ages two hundred. Instead, since the durations experienced are the same, so too will be the extended, quantifiable times each observer measures for himself: "they will have to coincide not only with respect to the different modes of *quantity* but even, if I may so express myself, in respect to *quality* for their inner lives have become indistinguishable, quite like their measurable features" (59).

Defenders of Bergson, following Deleuze, excuse his misunderstanding of special relativity by arguing that his interlocutors miss his point. It is a matter of ontology, not psychology, they argue.[10] The physicist, they maintain, still spatializes time, treating it as the counting of actual instants, but this clock time depends on lived duration and its virtual continuity; Einstein, mistaking duration for a merely psychological time, has thus confused actual, discrete, quantitative multiplicities with virtual, continuous, qualitative ones (see Deleuze 1991, 78–85; also Ansell Pearson 2002, 58–65; and Durie's introduction in Bergson 1999, xvi–xvii, xix–xx).[11] But this apology evades the ontological question of whether quantity can be reduced to an abstraction of quality or an external perspective on quality. By this principle, Bergson dismisses the bending of space and slowing of time as images used to support abstractions (1999, 20), affirms and rejects the reality of constitutive vibrations as convenient for his affirmation or denial of variable speeds of duration, and limits discontinuity to the extended and spatial. When Einstein responds to Bergson that "there is . . . no philosopher's time; there is only a psychological time that differs from the time of the physicist"

---

10. Deleuze (1991, 116) holds that Bergson intends not to oppose psychology to physics but to provide the metaphysics needed by relativity physics.

11. Scott (2006) provides a similar defense of Bergson, although it is oriented to portraying Bergson's criticisms as precursors to Heidegger's. Murphy (1999) allies Bergson's criticisms of relativity with those of quantum physics, finding parallels between Bergson's idea of a single duration and the real simultaneities implied by the phenomenon of nonlocality. Nevertheless, quantum theory presents obvious difficulties for Bergsonian continuity, as Bachelard demonstrates.

(Einstein, quoted in Bergson 1999, 159), he too goes beyond physics and psychology. The philosopher's attempts, he says, to derive physical time and the objective simultaneity of events from an agreed, subjective simultaneity, "did not, for a long time, lead to any contradiction because of the great speed of light" (159). But relativity exposes an essential discontinuity between the two times, and so the irreducibility of one to the other. The result: quantity does not rest on quality; one is not relative and the other absolute.

Deleuze himself holds that through the idea of vibrations Bergson reintroduces quantity into quality as a kind of virtual coexistence within quality. This virtual quantity, which Deleuze conceives as "degrees of difference" as against "differences of degree," is akin to what, in *Nietzsche and Philosophy*, he calls "difference in quantity" (see Deleuze 1983, 42–44) and is crucial, as will be seen, to Deleuze's deployment of Nietzsche against Hegelian dialectics. Yet Deleuze provides at best indirect textual evidence from Bergson to support his reading.[12] And indeed, although he would have qualms with the lack, lacuna, and negativity of Bachelard's dialectical reply to Bergson, Deleuze also breaks with Bergson on the same issues of continuity, the time of thought, and the production of the new.[13] The development from the movement-image to the time-image in Deleuze's cinema works (1986 and 1989) shows clearly the limits he finds in Bergson's philosophy. The movement-image, Deleuze explains, connects part to whole. A counted time perceived through movement is used to measure movement, but local motion also points beyond itself to an image of duration as an ever-changing open whole. Duration is thereby a second level of movement or a movement of movements: "the essence of the cinematographic movement-image lies in extracting from vehicles or moving bodies the movement which is their common substance, or extracting from movements the mobility which is their essence" (1986, 23). However, when time is read off motion, Deleuze contends, we are given only an indirect image of what it is (1989, 34–35). In

12. No textual references are provided in the key pages of *Bergsonism* (Deleuze 1991, 92–94). In his early essay "Bergson's Conception of Difference" (1999), Deleuze proposes that "one of Bergson's most curious ideas is that difference itself has a number, a virtual number, a sort of numbering number" (44); yet the subsequent presentation of degrees of difference fails to reintroduce number or quantity. Moreover, the link Deleuze proposes between vibration and relaxation (56) reinforces the opposite idea, that quantitative vibrations are secondary, merely the most relaxed form of qualitative internal difference that is contracted into a concrete but virtual whole (54–55).

13. Williams (2005) highlights some of these qualms but still holds Deleuze to affirm a virtual continuity against Bachelard. This reading of Deleuze's virtual seems unsustainable, however, when concepts such as the differenciator—which goes unmentioned in Williams's (2003) in-depth reading of *Difference and Repetition*—are introduced.

the transition to the direct time-image that defines modern cinema, following the exhaustion of the forms of perception, affection, and action given by movement, "time is no longer the measure of movement but movement is the perspective of time" (22). Time now becomes an unchanging form of what moves or changes. The direct time-image of Bergsonian duration is that of the coexistence of virtual past and actual present. Just as duration's movement-image goes beyond the local, empirical movements that it grounds, its direct time-image exceeds both the empirical succession of instants and the preservation of the past, providing their transcendental ground (98). But although plurality and heterogeneity appear in the two forms associated with this time-image—the sheets of the past and the peaks of the present (see 1989, chapter 5)—it remains parasitically attached to movement, being derived as a condition of possibility. Moreover, Bergson's image "essentially concerned the *order of time,* that is, the coexistence of relations or the simultaneity of the elements internal to time" (155). It does not go fully beyond the image of time it criticizes.

In contrast, Deleuze associates another time-image with Nietzsche.[14] It "concerns the *series of time,* which brings together the before and the after in a becoming, instead of separating them; its paradox is to introduce an enduring interval in the moment itself" (1989, 155). This image affirms a "power of the false" absent in Bergson, a power that "poses the simultaneity of incompossible presents, or the coexistence of not-necessarily true pasts" (131). It grounds neither temporal succession, the preservation of the past, nor empirical movement; it instead ungrounds them by posing a "false movement" (143), which, as such, is not really movement at all. Deleuze elsewhere calls this ungrounding a "line of flight" or a "becoming-minor" and he associates it with the production of the new, "of the intrinsic quality of that which becomes in time" (275). The form of the Nietzschean time-image is an "irrational cut" or "interstice" (277) bringing together incommensurables within an instant understood as a pure event. It replaces the open whole of duration with an outside or unthought that compels thinking: "this time-image puts thought into contact with an unthought, the unsummonable, the inexplicable, the undecidable, the incommensurable. The outside or the obverse of the images has replaced the whole, at the same time as the interstice or the cut has replaced association" (214).

14. Deleuze's turn from Bergson to Nietzsche is often noted but treated as a deployment of Nietzsche to deepen Bergson or as a Bergsonization of Nietzsche, rather than as a break with Bergson. See Ansell Pearson (2002, chapter 7), Borradori (2001), Boundas (1996), and Moulard (2002).

Bachelard and Deleuze offer two ways to break Bergsonian continuity—
which in neither case suggests an abandonment of Bergson but rather "a
renewal or an extension of his project" (Deleuze 1991, 115). The choice is
between two sorts of discontinuity—Bachelard's Hegelian and dialectical
negativity or Deleuze's Nietzschean irrational cut—but both options follow
a Bergsonian aspiration to raze abstractions and provide an account of concrete time. For Bergson, these abstractions are found in the physicist's chronological time;[15] Bachelard and Deleuze find a lingering abstraction in the
thesis of continuity. At issue, therefore, in the choice between Bachelard
and Deleuze, as in the choice between Hegel and Nietzsche, is what kind
of discontinuous, microscopic difference can fulfill the requirements of a
concrete philosophy of time and an ontology of sense.

15. "We would, besides, have to distinguish between the standpoints of philosophy and
science; the former rather regards the concrete, all charged with quality, as the real; the latter
extracts or abstracts a certain aspect of things and retains only size or relation among sizes"
(Bergson 1999, 45).

5

Disguised Platonisms

IS PLATONISM DEFINED MORE by the theory of Forms or by the account of reminiscence that enables access to them? Both might seem necessary to escape the aporias of definition most prominent in the early dialogues and attain a positive ground for becoming. Yet they have clearly experienced very different fates. In contrast to reminiscence, even Plato himself, particularly in *Parmenides,* casts doubt upon the coherence, efficacy, and very existence of the Forms, especially those of substances as opposed to qualities (see *Parmenides* 130b–134e).[1] Moreover, as demonstrated principally in the *Republic,* an unavoidable transcendence undermines the entire theory. This quandary is not found in the Forms themselves, which, although located beyond the physical world, remain knowable through introspection. It resides rather in the Form of the Forms, the Good, which remains opaque because it is not an object of knowledge but rather the source and medium for knowledge, as light is the medium for vision. Just as the sun, as opposed to the light emanating from it, cannot become visible except by a more powerful light shining upon it, knowledge of this medium would require another medium, and so on ad infinitum. It is thus, strictly speaking, impossible to know the Good. It must remain a mysterious foundation, always already withdrawn, leaving no basis for the metaphysical order of the divided line—the order of Form, copy, and simulacrum—oriented by it (see *Republic* 506d–511).

As the subsequent history of philosophy demonstrates, the Platonist aspiration for a foundation need not involve a "real world" transcending the world of becoming. The notion of a mysterious, unrepresentable "some-

1. All references to Plato's dialogues are taken from Plato (1961) and cite the dialogue, where appropriate, and the Stephanus pagination.

thing" need not take the form of a Platonic Idea and, as will be seen, is crucial to a philosophy of immanence. Compared to the Forms, variants of reminiscence have had more staying power, being used prominently by philosophers who explicitly pronounce their anti-Platonism but frequently identify Platonism primarily and narrowly in terms of transcendence. Examples include Bergson, who maintains that Plato's static Good, lacking any motivating force, is a mere ornament to the morality of obligation (see Bergson 1956, 87–88, 270), Bergson himself turning to a modified conception of Spinoza's *natura naturans* (58) and to his own concept of creative duration accessed through intuition and memory;[2] and Lacan, who proposes a recollection that "is not Platonic reminiscence—it is not the return of a form, an imprint, a [*sic*] *eidos* of beauty and good, a supreme truth, coming to us from the beyond" (Lacan 1981, 47). Against Plato's use of reminiscence as a link to unchanging eternity, both Lacan and Bergson deploy their versions to delineate processes within the infrastructures of chronological time. For Lacan these are the primary processes of the unconscious and the signifying synchrony of language, which reside in the gap between perception and consciousness (45–46, 56), while for Bergson they are the actualizations of the virtual past that take place in the delay effected by consciousness between perception and action (Bergson 1991, 65). Nevertheless, while there remain fundamental differences in their treatment of the past and memory, Lacan and Bergson agree in two important respects, which both have Platonic overtones: reminiscence does not involve the empirical recall of past perceptions or experiences, and it grounds becoming, action, and even the order of chronological time itself. Given such convergences, it remains unclear, under a broader definition of Platonism, whether Lacan and Bergson on crucial points do anything more than simply preserve and repeat aspects of Plato's foundationalism that seem sustainable despite the loss of traditional metaphysical guarantees and transcendences.

The Platonist motivation to firmly separate philosophy from sophistry and Plato's ultimate failure to distinguish Socrates from the sophists are well known (see, for example, Deleuze 1990, 253–66; Derrida 1981). Preliminary steps must be taken, however, before the attempt to draw these

---

2. Incredibly, Bergson (1956, 262–64) holds that Plato uses no argument from memory to demonstrate the soul's immortality but instead relies on a priori definition, and then proposes an alternative argument for such immortality based on the experience of memory independent of the body and on mystic intuition. For Plato's argument that recollection demonstrates the soul's independence from and existence prior to the body, see, of course, *Phaedo* 72e. The case for immortality is completed later in the dialogue through appeal to the mythical story of the soul's journey in the afterlife.

ultimately ambiguous battle lines can even be made. These steps establish "that becoming in general takes place with a view to being in general" (*Philebus* 54c). Things of this world, Plato notes, partake in a certain dualism. On the one hand, they participate in opposing qualities due to their relations to one another—"when you say that Simmias is taller than Socrates but shorter than Phaedo, you mean that at that moment there are in Simmias both tallness and shortness" (*Phaedo* 102b)—and due to their transience (see *Republic* 479a–b). On the other hand, "that which has a nature relative to self will retain also the nature of its object," such that "if hearing hears itself, it must hear a voice. . . . And sight . . . if it sees itself must have a color," while "that which is more than itself will also be less, and that which is heavier will also be lighter, and that which is older will also be younger" (*Charmides* 168d–e). In *Parmenides,* this thinking is extended beyond ascribed qualities to processes of becoming. Where one thing is older than another, for example, as they continue through time the older becomes relatively younger than the other without ever being younger, and vice versa; yet there is also no such becoming, since the difference in age remains the same throughout (*Parmenides* 154a–155c). In relations to self, becoming is rendered truly paradoxical.

> Whatever occupies time must always be becoming older than itself, and "older" always means older than something younger. Consequently, whatever is becoming older than itself, if it is to have something *than* which it is becoming older, must also be at the same time becoming younger than itself. . . . If one thing is already different from another, there is no question of its becoming different; either they both are now, or they both have been, or they both will be, different. But if one is in process of becoming different, you cannot say that the other has been, or will be, or as yet is, different; it can only be in process of becoming different. Now the difference signified by "older" is always a difference from something younger. Consequently, what is becoming older than itself must also at the same time be becoming younger than itself. Now, in the process of becoming it cannot take a longer or shorter time than itself; it must take the same time with itself, whether it is becoming, or is, or has been, or will be. So, it seems, any of the things that occupy time and have a temporal character must be of the same age as itself and also be becoming at once both older and younger than itself. (141a–d)

Existence in time thereby involves being at once in sync and out of sync with oneself. Becoming is characterized this way insofar as no end or purpose can exist within the process itself (155a), which further implies that a single direction of change can be given only after the fact or from the outside. This out-of-sync structure might even be attributable to time, had Plato not linked time to the creation and ordered movement of the heavens (*Timaeus* 37c–38b; see also *Statesman* 269d–270e), a move that leads directly to the problem, also faced by Augustine, of what occurred *before* time's existence (see *Timaeus* 39e; and Augustine 1961, book 12). Regardless of the last point, pure becoming is independent of the existence or nonexistence of any enduring principle or Form: "whether there is or is not a one, both that one and the others alike are and are not, and appear and do not appear to be, all manner of things in all manner of ways, with respect to themselves and to one another" (*Parmenides* 166b).

Plato, however, is largely uninterested in pursuing the implications of this line of thought. In *Philebus,* Socrates dismisses Protarchus's formulation of the duality of physical things as inappropriate to the problem of the one and the many, calling it "childish, obvious, and a great nuisance to argument" (14d; see 14d–15a generally). In the *Republic,* where justice in the city and soul is defined as the ordered harmony of parts, duality is denied when deriving the distinctive parts of the soul, although it is granted when arguing for the existence of Forms.[3] Although *Timaeus* distinguishes "that which always is and has no becoming, and . . . that which is always becoming and never is" (27d; see also 38b), it equates becoming with what is created in accordance with a cause (28a), not with the primordial chaos of uncreated matter, which, preexisting the ordered heavens and time and lacking any proportion except by accident (52d–53c, 69b), comes closer to the notion of pure becoming. Given that discourse and philosophy disintegrate without the fixed points provided by Forms (see *Parmenides* 133c–d), the problem of the one and the many must be posed through the intermediary of an ordered, numbered plurality: "we are not to apply the character of unlimitedness to our plurality until we have discerned the total number of forms the thing in question has intermediate between its one and its unlim-

---

3. Compare "it is obvious that the same thing will never do or suffer opposites in the same respect in relation to the same thing and at the same time. So that if ever we find these contradictions in the functions of the mind we shall know that it was not the same thing functioning but a plurality" (*Republic* 436b), with "and likewise of the great and the small things, the light and the heavy things—will they admit these predicates any more than their opposites? . . . No, he said, each of them will always hold of, partake of, both" (479b).

ited number . . . it is the recognition of those intermediates that makes all the difference between a philosophical and a contentious discussion" (*Philebus* 16d, 17a; see 15b–18b generally). This dialectical organization of the one and its limited plurality must precede any coherent move to the unlimited and dispersed; conversely, the unlimited can be grounded in unity only by the mediation of the limited many.[4]

Reference to this intermediary, however, also substantially alters the nature of unlimited becoming. Plato's post-*Parmenides* development should be seen neither in terms of progressive dissatisfaction with the Forms nor as a direct revision of the theory,[5] but rather as a shift to explore the theory's necessary conditions. In *Theaetetus*, which erects a strict separation between becoming and being and declares it wrong to say of a thing either that it is or that it is just one thing (152d–e, 157a–b) because it is really "an assemblage of many" (157b), knowledge remains undefined because it is impossible to specify false judgment as, for example, thinking something that "is not" (187d–189d). Conversely, in the *Sophist*, defining the false philosopher's being through the separation of the copy, which imitates the original, from the semblance, which merely appears to imitate (235d–236e), necessitates "that what is not, in some respect has being, and conversely that what is, in a way is not" (241d). In the *Statesman* (283d–284c), this need is linked directly to the existence of fixed norms that enable absolute and not simply relative measure: statecraft, requiring that excess and deficiency to be measured against the standard of the good man, calls for the existence of that which is not. Through these moves, the Forms' power is linked to isolating the copy from the simulacrum, granting existence to nonbeing, and mediating being and becoming through a limited plurality. A foundation in being is thus possible only through a domestication of pure becoming that places it within the aegis of being and defines and mediates even its absolute excesses and deficiencies through number and order. All this has the cost of conceding being and efficacy to the sophist and his semblances. But it is necessary to defeat the sophist's defense that he cannot be condemned for producing nonbeings because these can exist neither in reality nor in lan-

---

4. "When you have got your 'one,' you remember, whatever it may be, you must not immediately turn your eyes to the unlimited, but to a number; now the same applies when it is the unlimited that you are compelled to start with. You must not immediately turn your eyes to the one, but must discern this or that number embracing the multitude, whatever it may be; reaching the one must be the last step of all" (*Philebus* 18a–b).

5. On the issue of whether the later dialogues reflect an abandonment, revision, or reinforcement of the theory of Forms, see Adalier (2001), Allen (1965), Cornford (1935), and Kahn (2007).

guage (*Sophist* 236e–239d, 260a–261b) and to place the sophist within the purview of the transcendent truths that delineate his subordinate status. Establishing true being thus goes hand in hand with acknowledging the power of the false.

The Forms are accessible only after this ordering and, conversely, their grounding function presupposes this order. Without it, the argument in *Parmenides* that the worlds of being and becoming cannot connect to each other remains insurmountable.[6] When the demiurge imposes order and number on chaos (*Timaeus,* 53b), he creates a moving imitation of a static model (30c–d). Becoming's domestication is found here in the organization of time, which, through its dimensions of past, present, and future, "imitates eternity and revolves according to a law of number" (38a), allowing temporal things to copy eternity through repetition and propagation (*Symposium* 206e–208b). Reminiscence too depends on this order of time, which connects memory specifically to the past, its basic premise being "that what we call learning is really just recollection . . . what we recollect now we must have learned at some time before" (*Phaedo* 72e). Yet the role of reminiscence is not to recall past memories—an operation representing at best only its most generic sense—but rather to provide a route beyond time to the eternity that time copies. In other words, reminiscence, while being read off chronological time, does not actually concern this time, but rather its immemorial foundation—even though this foundation depends on an already ordered time. This explains the status of writing, which is condemned in *Phaedrus* (275a–d) as an impediment to living memory and true knowledge, serving as no more than a reminder and being unable to provide anything reliable and permanent (on this point see also Derrida 1981, esp. 95–117), but is praised in *Laws* (891a) for giving endurance to legislation. Writing can indeed preserve and remember, but only with things of this world; the truth of these things is accessed immediately through speech and dialectic. Hence Plato's own writings narrate verbal exchanges either directly or through later accounts from either participants (i.e., *Phaedo*) or those relay-

---

6. "Suppose, for instance, one of us is master or slave of another; he is not, of course, the slave of master itself, the essential master, nor, if he is a master, is he master of slave itself, the essential slave, but, being a man, is master or slave of another man, whereas mastership itself is what it is [mastership] of slavery itself, and slavery itself is slavery to mastership itself. The significance of things in our world is not with reference to things in that other world, nor have these their significance with reference to us, but, as I say, the things in that world are what they are with reference to one another and toward one another, and so likewise are the things in our world" (*Parmenides* 133d–134a).

ing the stories secondhand (i.e., *Symposium*). Even hearsay is legitimate, if transmitted orally.

As reference to transient physical exemplars cannot demonstrate what these instances share in common (see *Laches* 191d–192a; *Euthyphro* 6d–e; and *Greater Hippias* 288a) and as absolutes are not given to the senses (*Phaedo* 65d), knowledge of both physical truths and absolutes requires recollection of knowledge the soul had before embodiment.[7] This reminiscence tethers correct opinion to true knowledge (*Meno* 97e–98a). Although "to the highest and most important class of existents there are no corresponding visible resemblances, no work of nature clear for all to look upon" (*Statesman* 286a), recollection may still be triggered by exemplars that are either similar or dissimilar to the Forms (*Phaedo* 73e–74a). But Plato also maintains that the specific experience of beautiful things leads the soul toward Beauty itself, which is the Form of the Good (*Phaedrus* 250b–252b; *Symposium* 211a–212a). In the *Symposium* (206e), such recollection inspires love, "a longing not for the beautiful itself, but for the conception and generation that the beautiful effects." It is neither a yearning for one's other half, as suggested by Aristophanes (193a), nor even a desire for a quality or thing that one lacks, as suggested by Socrates (200a), since, according to Diotima's teaching, a middle position exists between fullness and lack, just as one exists between being and nonbeing (202a–206a). These desires are merely derivative, as genuine love "never longs for either the half or the whole of anything except the good" (205e). Nevertheless, desire's cravings refer to an experience of refilling and so to a dissonance between the body's sensation of deficiency and the soul's remembrance of fullness (*Philebus* 41c). Precisely because the body, which is lacking, cannot apprehend its refilling, desire's basis must be found in the soul, which "apprehends the replenishment, and does so obviously through memory" (35b). Desire, linked to reminiscence and the soul, thereby grounds knowledge, action, and becoming: "it is to the soul that all impulse and desire, and indeed the determining principle of the whole creature, belong" (35d).

In its structural function, Lacan's unconscious desire parallels Plato's on almost every major point. Desire, arising on the terrain of the Other, appears in the metonymic shifts of language—slips of the tongue, and so on—in which the conscious subject surprises itself (see Lacan 1981, 23–28).

---

7. "Thus the soul, since it is immortal and has been born many times, and has seen all things both here and in the other world, has learned everything that is. So we need not be surprised if it can recall the knowledge of virtue or anything else which, as we see, it once possessed . . . for seeking and learning are in fact nothing but recollection" (*Meno* 81c–d).

It is neither ontic nor ontological but primarily ethical, insofar as it relates to "the father, the Name-of-the-father, [which] sustains the structure of desire with the structure of the law" (34). Desire differs from pleasure, which finds its limit in completion and homeostasis (31); desire's limit, rather, resides in its object's impossibility, through which "it is sustained as such, crossing the threshold imposed by the pleasure principle" (31).[8] As a component of the unconscious, it "is inaccessible to contradiction, to spatio-temporal location and also to the function of time" and so it is rightly called indestructible (31–32). Nevertheless, desire does not actually lack temporal character but instead represents the enduring or permanent side of time's structure, comparable to the permanence of substance that for Kant underpins all temporal relations (see Kant 1965, A182–84/B224–27).[9] Desire and the unconscious more generally lack the transcendent character of Plato's eternity, but assume its function, transposing it onto a "pre-ontological" realm (see Lacan 1981, 29–30).

Recollection is again the mechanism to access, albeit incompletely, time's substructure. Its form is that of return or repetition (Lacan 1981, 48), but "repetition is not reproduction" (50). It is instead linked to hauling or towing, "to a *hauling* of the subject, who always drags his thing into a certain path that he cannot get out of" (51). It is a matter of history—"only psychoanalysis allows us to differentiate within memory the function of recollection . . . it resolves the Platonic aporias of reminiscence through the ascendancy of history in man" (Lacan 1977, 167)—but in a special manner. What is hauled along is the traumatic encounter with the Real, which constitutes subjectivity and introduces the subject to language. It is a "historical" event by virtue of referring to conditions of emergence that, in the terms given by a preexisting order of chronological time, must be located in the past. But this "does not mean that the facts . . . are purely accidental, or simply factitious, and that their ultimate value is reducible to the brute aspect of the trauma" (51). The recollection that fills in the gaps of the subject's history must not be confused with Bergson's "naturalistic inadequacy" (28) and his "myth of a restoration of duration in which the authenticity of each instant would be destroyed if it did not sum up the modulation of all the

---

8. Thus the *objet a* is not the aim of desire but "either a phantasy that is in reality the *support* of desire, or a lure" (Lacan 1981, 186).

9. "If indestructible desire escapes from time, to what register does it belong in the order of things? For what is a thing, if not that which endures, in an identical state, for a certain time? Is not this the place to distinguish in addition to duration, the substance of things, another mode of time—a logical time?" (Lacan 1981, 32). On the character of logical time see Lacan (1981, 39–40; 2006, 161–75).

preceding ones" (47–48). Recollection here is "not a question of biological memory, nor of its intuitionist mystification." Indeed, "it is not a question of reality, but of truth" (48) in that "it is the truth of what this desire has been in his history that the patient cries out through his symptom" (167). Through the spoken dialogue of the analytic setting—a dialogue that may be "made up of lies" (47)—recollection ascribes sense to the past, its effect being "to reorder past contingencies by conferring on them the sense of necessities to come" (48). This hauling along of the past is thereby the historization that gives sense to the subject's history: "What we teach the subject to recognize as his unconscious is his history—that is to say, we help him to perfect the present historization of the facts that have already determined a certain number of the historical 'turning-points' in his existence" (52). Hence, "its operations are those of history, in so far as history constitutes the emergence of truth in the real" (49), but this truth expresses "the limit of the historical function of the subject" (103). This limit, in turn, "represents the past in its real form, that is to say, not the physical past whose existence is abolished, nor the epic past as it has become perfected in the work of memory, nor the historic past in which man finds the guarantor of his future, but the past which reveals itself reversed in repetition" (103).

In this way, the Law of the Father functions as a mysterious ground. The status of its signifier, the phallus, as a simulacrum rather than an iron-hard reality or essence, does nothing to alter this role,[10] nor does the fact that "repetition demands the new" (Lacan 1981, 61), insofar as this novelty is explained only negatively.[11] The revelation of the traumatic Law "presents

10. "In Freudian doctrine, the phallus is not a phantasy, if by that we mean an imaginary effect. Nor is it as such an object (part-, internal, good, bad, etc.) in the sense that this term tends to accentuate the reality pertaining in a relation. It is even less the organ, penis or clitoris, that it symbolizes. And it is not without reason that Freud used the reference to the simulacrum that it represented for the Ancients" (Lacan 1977, 285).

11. Repetition "is turned towards the ludic, which finds its dimension in the new" and every repetition of the same thereby "conceals what is the true secret of the ludic, namely, the most radical diversity constituted by repetition in itself" (Lacan 1981, 61). However, Lacan explains this as a novelty by default, resting it on the failure of signification to fully accomplish its aims and the need to continually camouflage this failure: "It can be seen in the child, in his first movement, at the moment when he is formed as a human being, manifesting himself as an insistence that the story should always be the same, that its recounted realization should be ritualized, that is to say, textually the same. This requirement of a distinct consistency in the details of its telling signifies that the realization of the signifier will never be able to be careful enough in its memorization to succeed in designating the primacy of the significance as such. To develop it by varying the significations is, therefore, it would seem, to elude it. This variation makes one forget the aim of the significance by

us with the birth of truth in speech, and thereby brings us up against the reality of what is neither true nor false" (Lacan 1977, 47). As such, "there is nothing false about the Law itself, or about him who assumes its authority" (311), although all claims to be the Law's author—that is, its ground—are fallacious (310–11). Recollection's true role in analysis is found in the calling that constitutes the subject (Lacan 1981, 47). The subject, lacking an identity, cannot "think himself exhausted by his *cogito*" (Lacan 1977, 317) and is called onto the domain of the Other. At the same time, the nature of the trauma suggests a prior time when the subject was whole. The Word of the Father, in turn, both institutes the order of language and circulates a "Great Debt" (67) through the symbolic system, forming the basis of morality. This enigmatic ground institutes a series of substitutions and repetitions: substitutions of signifiers that replace the original, repressed signifier, so that "at each stage in the life of the subject, something always arrived to reshape the value of the determining index represented by this original signifier (Lacan 1981, 251); and repetitions of love, whereby the subject, seemingly searching for a sexual complement, in fact seeks "the part of himself, lost forever, that is constituted by the fact that he is only a sexed living being, and that he is no longer immortal" (205).

In Bergson, the foundation of morality goes beyond the form of law, but it remains grounded in the past. *The Two Sources of Morality and Religion* has an obvious connection to *Creative Evolution*—the latter, Bergson (1956, 249, 256–57) says, identifies the *élan vital* as the impetus for evolution but does not reach the source of this impulse—and here Bergson certainly turns to transcendence. But there is a no less important relation, with comparable implications, between the two forms of religion and the two forms of memory in *Matter and Memory*.[12] Static religion, expressed as social obligation in a closed society, corresponds to motor or habit memory, which acts the past automatically without recalling it (Bergson 1991, 78–82). Obligation must be conceived "as weighing on the will like a habit, each obligation dragging behind it the accumulated mass of the others" (Bergson 1956, 25), and it takes the form of a moral or rational imperative only post hoc, when consciousness demands reasons and philosophical justifications (25–26, 91).

---

transforming its act into a game, and giving it certain outlets that go some way to satisfying the pleasure principle" (61–62).

12. Lawlor (2003) also connects these two texts even while arguing that Bergson both "reverses" and "twists free" of Platonism. Nevertheless, with respect to *The Two Sources*, Lawlor acknowledges: "Perhaps in the final analysis we have to characterize Bergson as a philosopher of transcendence rather than as a philosopher of immanence" (xii).

Moral habits are established by repetitive practices and ceremonies (201), just as lessons must be read many times to be learned by heart (Bergson 1991, 79–82). The final product of obligation resembles and even substitutes for instinctual habits: "obligation is to necessity what habit is to nature" (Bergson 1956, 14); "no one obligation being instinctive, obligation as a whole *would have been* instinct if human societies were not, so to speak, ballasted with variability and intelligence" (28). Static morality thereby acts as a kind of social ego overlaying the individual's ego, "because his memory and his imagination live on what society has implanted in them" (15). It stabilizes individuals "who cannot find within themselves the resources of a deep inner life" (16).

Seemingly at the base of this habit morality is a buried memory of prohibition—"the remembrance of forbidden fruit is the earliest thing in the memory of each of us, as it is in that of mankind"—which is enforced during youth by authority figures such as parents and teachers, who "seemed to act by proxy" (Bergson 1956, 9). Nevertheless, this apparent origin of morality is a mirage. This is not simply because the memory of prohibition can be mythical, its function being simply to police the excesses of intellect: "Since instinct no longer exists except as a mere vestige or virtuality . . . it must arouse an illusory perception, or at least a counterfeit of recollection so clear and striking that intelligence will come to a decision accordingly" (122). Even granting this point, this account "gives us only a figurative symbolization of what actually occurs" (122). Automatic habit memory is only a terminal form, one that reduces the past to the present: "the lesson once learned bears upon it no mark which betrays its origin and classes it in the past; it is part of my present, exactly like my habit of walking or of writing" (Bergson 1991, 81). Just as habit memory finds its foundation in genuine recollection memory (83–84), habit morality is a mere abstraction when taken in isolation and finds its conditions of emergence in a very different activity: it "becomes, in comparison with the other [dynamic morality], something like a snapshot view of movement" (Bergson 1956, 59).

Dynamic religion, then, expressed as a universalism surpassing all closed societies of the first moral system, corresponds to the energy associated with the automatic, complete, and ordered recording of memory-images, the second form of memory. The latter may be spontaneous, in which case it remains vague and ever present, or voluntary, in which case vagueness is suppressed in an actualized recollection (Bergson 1991, 86–87). In both cases, however, memory-images refer to and partake in pure memory (133) and, therefore, the activity of duration, which is itself read off the order of time given in habit.[13] Pure memory underpins the mechanisms of bodily habit, while the body apparatus furnishes the means for the virtual past to become actual (152–53). The reflux of the past into the present, being a qualitative and simple change, is a real movement that becomes relative and discontinuous only when subsequently translated into quantity (195–96). Similarly, static morality finds its real foundation in dynamic morality's creative emotion, just as the immobile is derived from the mobile but not vice versa (Bergson 1956, 58). This emotion's inspirational force travels not through any divisible space or time but by a simple, qualitative leap (53–54). Nevertheless, "religious dynamism needs static religion for its expression and diffusion" (179). Inspiration first lodges itself in the existing static religion in order to supplant it (216–17), and then, as its emotion dies down, coagulates into new formulae of social obligation (49–50) and their subsequent intellectual and philosophical justifications. Dynamic morality must be incarnated in exceptional individuals, who take up the remnants of earlier impulses of inspiration and continue them forward: "What the mystic finds waiting for him, then, is a humanity which has been prepared to listen to his message by other mystics invisible and present in the religion which is actually taught" (239). Like the activity of *élan vital*, these figures consolidate the impulse from the past and propel it into the future, where it is again taken up, developed to a higher intellectual level, and pushed onward: "We do, as a matter of fact, see a first wave, purely Dionysiac, merging into Orphism, which was of a higher intellectual character; a second wave, which we might call Orphic, led to Pythagoreanism, that is to say, to a distinct philosophy; in its turn Pythagoreanism transmitted something of its spirit to Platonism, and the latter, having adopted it, in time expanded naturally into Alexandrine mysticism" (220). The past, then, is the energetic foundation for the establishment and development of both moralities.

The rather obvious residues of transcendence in Bergson and Lacan lie in the former's appeals to mysticism and the latter's use of the language of negative theology with respect to the phallus and the feminine. In these moments, Platonism does not seem very far away. But each thinker also

---

13. "No doubt there is an ideal present—a pure conception, the indivisible limit which separates past from future. But the real, concrete, live present—that of which I speak when I speak of my present perception—that present necessarily occupies a duration. Where then is this duration placed? Is it on the nearer or on the further side of the mathematical point which I determine ideally when I think of the present instant? Quite evidently, it is both on this side and on that, and what I call 'my present' has one foot in my past and another in my future" (Bergson 1991, 137–38).

disguises a more subtle Platonism within an incomplete move to immanence. As has been suggested earlier with respect to Bergson, and as will be
developed later with respect to Bergson and Lacan, continuing abstractions
remain in their conceptualizations of difference, comparable to those of
Hegelian logical contradiction, which render improper or incomplete syntheses. Variants of Platonic reminiscence, defined by reference to the chronological order of time, are thereby able to survive and continue to play a
foundational role not because of a full return to transcendent Forms but
because immanence has been completed only in abstraction. One effect is
the repetition of rather familiar moral and political claims. Lacan, parallel-
ing Plato's claim in the *Statesman* (293c–301a) that in the absence of an
ideal society that can afford to be lawless, strict adherence to the law is
paramount, links the problem of today's barbarism to the decline of the
superego and the ego ideal, and so to the decline of the Law (Lacan 1977,
26–27). When assessing the "faith, so difficult to sustain," that allows Spinoza to detach from human desire, he simply concludes: "Experience shows
us that Kant is more true, and I have proved that his theory of consciousness
. . . is sustained only by giving a specification of the moral law which, looked
at more closely, is simply desire in its pure state" (Lacan 1981, 275). Morality, in short, must take the form of law. Bergson, of course, insists on another route to alleviate what he sees as modern society's love of luxury and
corresponding moral decline (see Bergson 1956, 298–312), yet his alternative parallels another Platonic track that leads beyond the Law to the Idea:
"The generality of the one [static morality] consists in the universal acceptance of a law, that of the other [dynamic morality] in a common imitation
of a model" (34). If abstract immanence camouflages continuing Platonisms
in this way, then the elimination of this abstraction through a conception of
concrete difference renders a full reversal of Platonism. This difference is
linked to a structure of time that goes beyond what grounds a chronological
temporal order. It is further linked, as will be seen, to a different ethical
imperative, which Nietzsche calls the revaluation of values.

6

Syntheses of Difference and Contradiction

IN SUGGESTING A ROUTE from Hegelian dialectics, Deleuze asks, "Is not contradiction itself only the phenomenal and anthropological aspect of difference?" (Deleuze 1997b, 195). This might suggest a simple inversion of the historical and logical that, as with Marx, would see our logic as the product of our history, not the reverse. But Deleuze refuses this route: a historical dialectic, for him, advances no further than a logical or ontological dialectic in terms of finding a conception of difference adequate to a philosophy of immanence. What is required is a concept of becoming that exceeds and gives sense to the becoming of history,[1] just as Hegel's logos becomes in a nonhistorical manner while underpinning the sense of history. Hegel's, however, is a becoming of contradiction, which negates itself into the phenomenological passage of history. Deleuze's, by contrast, is a movement of difference surpassing contradiction—in Nietzschean terms, also employed by Deleuze, it is *untimely* or in the time of the eternal return—although it too implies a temporal or temporalizing becoming that is not historical. Hyppolite's reading of Hegelian contradiction specifies the requirements Deleuze must satisfy for his alternative ontology of sense. Hegel's conceptions of force, quantity, and quality provide the counterpoint for Deleuze to develop his concept of difference.

As Hyppolite maintains, only speculative contradiction, for Hegel, can fully determine identity and meaning so as to provide sense. Merely empirical differences and transcendental contradictions that fall short of specula-

---

1. "The thing is, I became more and more aware of the possibility of distinguishing between becoming and history. It was Nietzsche who said that nothing important is ever free from a 'nonhistorical cloud.' This isn't to oppose eternal and historical, or contemplation and action: Nietzsche is talking about the way things happen, about events themselves or becoming. What history grasps in an event is the way it's actualized in particular circumstances; the event's becoming is beyond the scope of history" (Deleuze 1995, 170).

tive reflection ultimately establish *indifference* and reinstate a philosophy of essence. Empirical thought denies the power of determination to negative difference: to say what a thing is not tells us nothing of what it is; instead, knowledge requires a positive content. This position, however, contradicts itself, since any positive content, taking the form, "X is Y," refers the thing beyond itself, so that it both is and is not itself.[2] Kant recognizes this contradiction and the totality that follows from the synthetic character of understanding and judgment, but he fails to appreciate its full implications. Kant reduces the Absolute to an Idea posited by thought as its condition and limit, failing to surpass the understanding's separation of subject and object and falling back onto psychologism and anthropomorphism (Hyppolite 1997, 82–83). Both Kantian and empirical thought thus carry residues of indifferent positivity—empirical diversity for one and the noumenal thing-in-itself for the other. Speculative knowledge surpasses these essentialisms, showing that no essence lies behind appearance because the Absolute is mediation.

Speculative difference, however, must take the form of contradiction. The Absolute can express itself only by sustaining its unity through diverse forms; to be self-determining, it must distinguish itself from its opposite without becoming one pole of this opposition. Negation must be compatible with identity, and opposition alone sustains both genuine diversity and identity, because a thing is individual only by differing from *everything* it is not: "Opposition is inevitable . . . because each is in relation with the others, or rather with all the others, so that its distinction is its distinction from *all the rest*" (Hyppolite 1997, 115). Negation must also inhabit both subject and object, thought and existence. Only then can speculative thought raise the Absolute from substance to subject, becoming the self-expression of being: "Speculative knowledge can be simultaneously knowledge of being and self-knowledge only because *to know oneself is to contradict oneself,* only because these two moments that we ordinarily separate in order to attribute one to the object, the other to the subject, truth and reflection, being and the self, are *identical.* Their identity in their contradiction is the very dialectic of the Absolute" (76).

Being real, negation cannot be limited to human thought or propositions.

---

2. "The rule of empirical knowledge lies in not contradicting itself in its object, and, since this rule is merely negative, the rule amounts to looking for the truth in the content, which is alone considered positive. But to say that A is B is already to contradict oneself, because this is to come out of the A in order to affirm something else about it; it is to say that it is not-A and not merely A" (Hyppolite 1997, 79).

Bergson, Hyppolite argues, errs in denying real negativity and holding the apparently equal status of positive and negative propositions to be illusory because negative propositions can only correct error and never determine being (108–11, 122–24). The second claim applies only to empirical propositions, and Bergson contradicts the first by admitting distinctions in nature, since "negation and distinction imply one another" (109). Once negation is attributed to being, contradiction alone can raise being to subjectivity. Spinoza and Leibniz, for Hyppolite, fail here: Spinoza's substance lacks self-reflection and produces external diversity, failing to mediate difference; Leibniz's monads incorporate self-reflection but remain in-themselves while referring to a transcendent God who creates them (150). Ultimately, contradiction proves to be the maximal form of difference. Lesser differences pass into it: "Hegelian dialectic will push . . . alterity up to contradiction. Negation belongs to things and to distinct determinations insofar as they are distinct. But that means that their apparent positivity turns out to be a real negativity. This negativity will condense the opposition in negation; negation will be the vital force of the dialectic of the real as well as that of logical dialectic" (113).

Although Deleuze employs Spinoza, Leibniz, and Bergson against Hegel, he affirms the basic principles underlying these Hegelian criticisms—that philosophy must be an ontology of sense and therefore an ontology of *internal* difference that leaves no indifferent remainder: "If philosophy is to have a positive and direct relation with things, it is only to the extent that it claims to grasp the thing itself in what it is, in its difference from all that it is not, which is to say in its *internal difference*" (Deleuze 1999, 42–43).[3] Unsurprisingly, Deleuze echoes Hegel by criticizing Spinoza's priority of substance over modes[4] and Leibniz's chess-playing God.[5] What, then, is Deleuze's re-

3. Hegel explains inner difference as follows: "we must eliminate the sensuous idea of fixing the differences in a different sustaining element; and this absolute Notion of the difference must be represented and understood purely as inner difference, a repulsion of the selfsame, as selfsame, from itself, and likeness of the unlike as unlike. We have to think pure change, or *think antithesis within the antithesis itself*, or *contradiction*. For in the difference which is an inner difference, the opposite is not merely *one of two*—if it were, it would simply *be*, without being an opposite—but it is the opposite of an opposite, or the other is itself immediately present in it" (Hegel 1977, §160).

4. "Spinoza's substance appears independent of the modes, while the modes are dependent on substance, but as though on something other than themselves" (Deleuze 1994, 40).

5. While using Leibniz to develop the conceptions of vice-diction and incompossibility against dialectical contradiction, Deleuze also holds Leibniz to have limited these ideas by relying on a transcendent God who chooses the best possible world according to a principle of maximum convergence or compossibility. See Deleuze (1994, 42–44, 45–51; 1990, 59, 110–12, 171–72, 259–60).

sponse to Hegel? Put simply: "if the objection that Bergson made against Platonism was that it stopped at a *still external conception of difference,* the objection that he makes to a dialectic of contradiction is that it remains with a *merely abstract conception of difference*" (53).[6] Contradiction may demonstrate the internal passage of a thing into its opposite, but it remains an abstract difference mediating abstract beings. Although this parallels Marx's critique that Hegel merely derives abstractions from abstractions, Deleuze aims to show not that Hegel has misplaced the dialectic in consciousness rather than labor, but that contradiction is necessarily less than difference rather than more.

> We are told that the Self is one (thesis) and it is multiple (antithesis), then it is the unity of the multiple (synthesis). Or else we are told that the One is already multiple, that Being passes into nonbeing and produces becoming. . . . To Bergson, it seems that in this type of *dialectical* method, one begins with concepts that, like baggy clothes, are much too big. The One in general, the multiple in general, nonbeing in general. . . . The concrete will never be attained by combining the inadequacy of one concept with the inadequacy of its opposite. (Deleuze 1991, 44)

More profoundly, contradiction and negativity are false problems "whose expression does not respect differences of nature" (Deleuze 1999, 46). They are "retrospective illusions" (53) and "merely external views of this internal difference" (49), but they also arise from this difference.[7] Hegelian contradiction, Deleuze argues, presupposes and is a symptom of difference, making the dialectic a superficial image of a more complex dynamic (Deleuze 1983, 156–59).

Deleuze does counterpoise dialectical mediation to immediate differentiation. He says that, "According to Hegel, the thing differs from itself because it differs in the first place from all that it is not, such that difference

---

6. Hardt (1993, 4–5, 7–8) seems to miss the distinction Deleuze draws between Platonist and Hegelian dialectics, holding Deleuze to admonish Hegelian contradiction for being an external conception of difference. Deleuze does locate a continuing externality in Hegelian dialectics, related to its treatment of difference as contradiction, which creates a gap between logic and phenomenology. However, this is not the same as holding that Hegelian contradiction posits an external other. Indeed, if Deleuze were to say this of Hegel, the critics would be correct to say that he completely misunderstands Hegel.

7. "We will see how this illusion is born, and what in turn grounds it in differences of nature themselves" (Deleuze 1999, 45–46).

goes to the point of contradiction" (Deleuze 1999, 53), seeming to miss that for Hegel a thing is *at the same time* also what it is not.[8] Moreover, he pits Nietzsche's affirmative forces against the dialectic's negative forces, holding that "in Nietzsche the essential relation of one force to another is never conceived of as a negative element in the essence. In its relation with the other the force which makes itself obeyed does not deny the other or that which it is not, it affirms its own difference and enjoys this difference" (Deleuze 1983, 8–9). These comments suggest a return to the immediacy and positivity Hegel easily dismisses and the reference to difference in the essence seems odd given Deleuze's acknowledgment that Hegel's is a philosophy of sense, not essence. However, Deleuze's line of thinking becomes clear in light of Hyppolite's reference to contradiction being essential difference because it defines the identity of a thing through opposition.[9] Deleuze does not seek a return to some predialectical immediacy, because such immediacy remains within a paradigm of identity; however, dialectical contradiction is the maximal form of difference only within this same paradigm (see Deleuze 1994, 49–50). Deleuze here comes close to Adorno's view that Hegelian negativity reaches identity only by already presupposing it: if dialectical negation results in positive identity, Adorno says, then it has not been negative enough.[10] Similarly, Deleuze maintains that contradiction fails to go far enough to reach difference. Immediate differentiation, then, refers to a difference that is greater because it exceeds any mediation that would make difference compatible with identity. Deleuze calls this excessive difference affirmative or positive, but this is not the positivity of an indifferent thing-in-itself criticized by Hegel, Adorno, and Deleuze. For both Adorno and Deleuze, what goes beyond contradiction is *nonidentity,* and what is more concrete than the Hegelian Identity of identity and difference is another form of synthesis: Adorno's nonidentity of identity and difference and Deleuze's disjunctive synthesis. The further task that Deleuze pursues is showing how difference and disjunction produce the conditions under

8. However, in *Difference and Repetition,* Deleuze writes of Hegelian contradiction: "Each contrary must further expel its other, therefore expel itself, and become the other it expels" (1994, 45).

9. "If identity suits things, dissimilarity or intrinsic difference also suits them, since they must be distinguished or differentiated in themselves from all the others. This difference (found within them) is essential difference, because it is the difference posited in the identity of the thing; this difference is what puts the thing in opposition to *all the rest*" (Hyppolite 1997, 119).

10. "To negate a negation does not bring about its reversal; it proves, rather, that the negation was not negative enough" (Adorno 1995, 159–60).

which they are mistaken for identity, contradiction, and dialectical mediation.

How, then, does Deleuze elaborate a rival to Hegel's conception of internal difference within these parameters? *Nietzsche and Philosophy* develops a subtle argument against Hegel using Nietzsche's conception of force. Nietzschean forces may appear to have little in common with those presented in Hegel's *Phenomenology* (1977, §§132–65), yet Deleuze warns against confusing them (Deleuze 1983, 8–10). Through Nietzsche, Deleuze poses an alternative to the negativity of Hegelian force, and in this way offers a rethinking of synthesis that breaks with dialectics.

For Hegel, "force" designates the movement from unity to multiplicity and back into which the object of perception dissolves. At stake is how the object can have meaning or sense given that it supposedly has its own content, yet it is defined through its properties, which relate it to others. As a substantive unity, a being-in-itself, the object relates to others, expressing itself in multiple properties; but as these properties interpenetrate and define a substantial whole, the plurality constitutes the unity. The notion of force sublates this contradiction within the perceived object by encompassing the moments of being-in-itself and being-for-another, performing the synthesis necessary for understanding, wherein the object, referring outside itself, nonetheless refers to itself alone. Initially, the being of force appears in its unity, while its expression is external to force, located in its multiple relations to others. But as force's expression is necessary, these others, though external, are also internal to force. Furthermore, they must themselves be forces, since they require their own substantiality in order to solicit expression.[11] This plurality of reciprocally determining, internally related forces reverses the priority of being-in-itself over being-for-another: since force attains unity only through its relations to other forces, unity becomes a mere moment in a more encompassing movement of being-for-self *through* being-for-another. Force is essentially relational, gaining specificity through its differences from other forces that are also identical to it, making these

---

11. "The subsistence of the unfolded 'matters' outside of Force is thus precluded and is something other than Force. Since it is necessary that *Force itself* be this *subsistence*, or that it *express* itself, its expression presents itself in this wise, that the said 'other' approaches *it* and solicits it. But, as a matter of fact, since its expression is *necessary*, what is posited as another essence is in Force itself. We must retract the assertion that Force is posited as *a One*, and that its essence is to express itself as an 'other' which approaches it externally. Force is rather itself this universal medium in which the moments subsist as 'matters'; or, in other words, Force *has expressed itself*, and what was supposed to be something else soliciting it is really Force itself" (Hegel 1977, §137).

relations of opposition or contradiction. Force sunders itself into its other, which defines it by opposing or negating it, appearing to be what the force is not; yet this opposite, by defining the identity of force, is immanent to it and therefore part of its identity. The synthetic being of force thereby total-izes itself, as anything outside of or opposed to this totality is always already part of it. The result is the Hegelian Absolute: the Identity of identity and difference. This mediation is the sense expressed in any object.

For Deleuze, Nietzsche's concept of force similarly displaces simple sub-stantive notions such as the ego, the thing-in-itself, the atom, or more gener-ally the object prior to its relations (Deleuze 1983, 6–8). Moreover, it is linked to meaning and sense: "We will never find the sense of something (of a human, a biological or even a physical phenomenon) if we do not know the force which appropriates the thing, which exploits it, which takes possession of it or is expressed in it" (3). However, opposition and contradic-tion being abstract, a more concrete form of force relations is needed. In *Nietzsche and Philosophy,* this is provided through a rethinking of quality and quantity. This move too is usefully framed in relation to Hegel.

In Hegel's logic, quantity is derived from quality, the quantitative unit being the dialectical synthesis of limit and alternation implied by qualitative or determinate being; but quantity also reinvokes quality, since a sufficient quantitative change becomes a qualitative change, resulting in the synthesis of quantity and quality as measure (Hegel 1975, §§89–111). This order of deduction from quality to quantity is significant, for it makes quantity an external modification of being that must be reintegrated dialectically. This reintegration is performed by repeating the moves of the dialectic of qual-ity—the equalization of quantitative units ("Quantity . . . has two sources: the exclusive unit, and the identification or equalization of these units" [§100]) and the rejection of the spurious infinity of extension (§§104–6). Quantity's external status allows Hegel to dismiss mechanism and atomism as abstract understandings that derive quality from external quantitative re-lations among purportedly independent entities (§§98–99, 195). The equal-ization of quantities drives a dialectical progression that makes all quantities and qualities measurable.

For Nietzsche too mechanism's "purely quantitative determination of forces remained abstract, incomplete and ambiguous" (Deleuze 1983, 43). However, the problem with mechanism is not just that it treats an external notion of quantity as the whole of reality. It is also found in mechanism's *equalization of quantities,* which is affirmed by Hegelian logic even in its critique of mechanism. In short, equality—the linchpin for dialectical medi-

ation in general and the mediation of quantity and quality in particular—
carries a problematic abstraction. Quantity is not an external modification
of being. Rather, quality, in the form of equality, is imposed from the outside onto quantity, annulling the difference germane to quantity.

Removing this abstraction reveals a new kind of quantity, designating an
internal rather than an external difference. Forces linked through unmediatable relations that differ from contradiction and the identity of contradictories are now said to differ quantitatively. Quantity—or, rather, difference
in quantity—is the internal difference that specifies forces, but this is neither a mechanistic nor a dialectical quantity. It is instead quantity *that cannot be equalized,* where differences in quantity cannot be measured by any
fixed scale. This internal difference maintains the heterogeneity of forces
that pass into one another, so that it is not a merely empirical diversity
failing to go as far as contradiction. Difference in quantity thereby denotes
another synthesis, in which forces are neither externally related things-in-
themselves nor moments within a dialectical synthesis of identity and opposition. Forces are rather part of a *disjunctive synthesis,* relating *through their
difference* rather than through identity. This synthesis, again, follows from
the elimination of abstract equality.

Quality is now an externalization of quantity or quantitative difference.
It arises with forces that gain meaning or sense only through internal but
unmediatable differences in quantity. Qualities are therefore heterogeneous, just like quantities, since they emerge from differences in quantity
that cannot be equalized.[12] The difference in quantity of related forces is a
difference in power: in unequal relations, one force necessarily dominates
while the other resists. This difference in power is internal, no force being
strong or weak in itself but only through its relations.[13] However, these are
not relations of simple inequality, but rather of *disequilibrium* or inequality

12. "Each time that Nietzsche criticises the concept of quantity we must take it to mean
that quantity as an abstract concept always and essentially tends towards an identification, an
equalisation of the unity that forms it and an annulment of difference in this unity. Nietzsche's reproach to every purely quantitative determination of forces is that it annuls, equalises
or compensates for differences in quantity. On the other hand, each time he criticises quality
we should take it to mean that qualities are nothing but the corresponding difference in
quantity between two forces whose relationship is presupposed. In short, Nietzsche is never
interested in the irreducibility of quantity to quality; or rather he is only interested in it
secondarily and as a symptom. What interests him primarily, from the standpoint of quantity
itself, is the fact that differences in quantity cannot be reduced to equality. Quality is distinct
from quantity but only because it is that aspect of quantity that cannot be equalised, that
cannot be equalised out in the difference between quantities" (Deleuze 1983, 43–44).

13. "But it should be kept in mind that 'strong' and 'weak' are relative concepts" (Nietzsche 1974, §118).

in flux, meaning that resisting forces can always overturn dominant ones. This is not a dialectical reversal that would resolve the heterogeneity of forces; if it were, Nietzsche would be reestablishing a simple hierarchy of dominating and dominated positions that, measuring inequality by a fixed scale of power, would remain as much an abstraction as the equality he challenges.

The quantitative difference among forces being one of relative strength and weakness, the corresponding qualities are active and reactive. Dominant forces are active, meaning that they command, create, transform, and overcome; dominated forces are reactive, meaning that they must work by adaptation, compromise, and utility (Deleuze 1983, 40–44). The active or reactive quality of force therefore indicates the tactics or means by which force exercises its power (54). Finally, any configuration of forces manifests a will to power, which is not a will to grasp power, although this desire for power regularly appears in certain forms of the will to power. The will to power, Deleuze says, is the principle of the quality of force and the signification of the sense of related forces (83, 85). It is what the configuration of active and reactive relations expresses. This expression is either affirmative or negative. What emerges from relations of strife is the will either to affirm strife or to deny it: "What a will wants, depending on its quality, is to affirm its difference or to deny what differs" (78). Affirmative and negative wills to power are closely related but not identical to active and reactive forces. Affirmation expresses active forces becoming dominant, while negation expresses forces in their becoming-reactive (54).

The will to power is also force's being-in-itself, the independence that, as with the dialectical movement of forces, is a moment within a more encompassing relation. It therefore has dual aspects, Deleuze argues, as a complement to and an internal factor of force, as a differential and a genetic element of forces, and as a product of force relations and the determinant of these relations (49–52).[14] Affirmative and negative wills to power, constituted by relations of domination and resistance, cannot be dialectically reconciled. The affirmative will to power of the strong is not the opposite or contradictory of the negative will to power of the weak, nor are these the two wills of Hegel's master and slave—indeed, only from the slave's perspective do they appear as opposites (10). The slavish perspective, which for Nietzsche and Deleuze arises from the nondialectical play of forces, pres-

14. See also Deleuze (1994, 125): "the eternal return is indeed the consequence of a difference which is originary, pure, synthetic and in-itself (which Nietzsche called will to power)."

ents the illusion or false problem of contradiction, but this makes it nothing more than a superficial image, a falsification of internal differences in quantity.

The slavish will to power stamps contradiction onto a world that is, morally and otherwise, much more complex and ambiguous. It initiates a moral reversal that takes what is good also to be pure, unchanging, and universal, and what is bad or evil to be the opposite of these. This reversal is driven by force relations themselves, as the force unable to assert itself must still express its will to power, and in its frustration with its weakness it wills an ideal in which forces are equalized or measured by fixed values. It matters little whether purity is subsequently mediated with its opposite, because opposition is the falsification that dialectics takes up and continues.[15] Opposition, contradiction, and their correlate, identity, constitute an inverted image of internal quantitative differences that cannot be reconciled, equalized, or fixed in an unambiguous hierarchy. This inverted image presents itself as universal sense (55–58). Only the weak, unable to act and thus viewing action as recipients or third parties, would initiate this reversal (73–74). That they are only recipients of action, however, implies another perspective and, consequently, another mode of thinking, being, and acting—that of the affirmative will to power.

It is often held that Deleuze cannot coherently distinguish this active, nondialectical perspective from the slavish universalization of identity and opposition. Deleuze himself highlights this difficulty, stating that the negative will to power alone is knowable, although it finds its essence in an affirmative will to power exceeding it.[16] Vincent Descombes argues that Deleuze must distinguish slave morality's opposition from noble morality's nondialectical difference, which allows the noble to affirm and enjoy his difference. But this, Descombes says, is impossible. For the noble seeking to differentiate rather than oppose himself to the slave, the slave's oppositional logic will appear as another nondialectical difference; conversely, for

15. "There would have been no need to put the dialectic back on its feet, nor 'to do' any form of dialectics if critique itself had not been standing on its head from the start" (Deleuze 1983, 89). Also, "the Hegelian dialectic is indeed a reflection on difference, but it inverts its image" (196).

16. "Thus nihilism, the will to nothingness, is not only a will to power, a quality of the will to power, but the *ratio cognoscendi of the will to power in general*. All known and knowable values are, by nature, values which derive from this *ratio*. . . . The other side of the will to power, the unknown side, the other quality of the will to power, the unknown quality, is affirmation. And affirmation, in turn, is not merely a will to power, a quality of the will to power, it is the *ratio essendi of the will to power in general*" (Deleuze 1983, 172–73).

the slave opposing himself to the noble, the noble's affirmation will appear as another negative opposition. From the perspective of both noble and slave, then, opposition and difference appear identical: "The non-identity of difference and opposition will appear to both as an *identity*" (Descombes 1980, 165). A further difficulty relates to the practice of noble affirmation: if he truly affirms difference, the noble must relate to the slave, lest he affirm only identity-in-itself; but if the noble defines himself in relation to the slave, this differentiation cannot appear different from opposition.

> The day that the Master with all his self-assurance meets, not another Master (i.e. another affirmation destined to negate him) but a Slave, he will learn the difference between a Master and a Slave, between a difference and an opposition. Thenceforth he will see that what he negates in the Slave is not another affirmation, but the actual negation of his own affirmation. He refuses the Slave's negation of him, but the discovery weakens him immediately. The time is coming when the Master, having discovered his own likeness in the Slave, will emancipate him. Indeed, how can it be distinguished, after a number of encounters between affirmation and negation, whether the *no* that one of the adversaries has just uttered precedes the *yes*, or follows it? (167)

This reading, however, misses many subtleties. It conflates power relations and the will to power, illicitly holding the existence of a nondialectical movement of forces to require the capacity of some will to recognize the difference between this movement and negative opposition. It treats noble and slavish wills as exhaustive alternatives, when clearly, for both Nietzsche and Deleuze, the Overman fits neither category neatly. Nietzsche himself acknowledges a falsification within noble morality, saying that the nobles mischaracterize the slaves even if the slaves mischaracterize the nobles more.[17] Ironically, the error in both cases is a reduction of difference to opposition: the slaves, unable to comprehend noble affirmation, compress it onto a moral schema of good and evil in order to attribute intentional harm to the nobles; the nobles characterize the slaves as their opposite—weak rather than strong—and fail to see the slaves as a threat. Ultimately, Descombes's

---

17. "Even supposing that the affect of contempt, of looking down from a superior height, *falsifies* the image of that which it despises, it will at any rate still be a much less serious falsification than that perpetrated on its opponent—*in effigie* of course—by the submerged hatred, the vengefulness of the impotent" (Nietzsche 1967, 1.10).

reading foists upon Deleuze the very alternative Deleuze considers spurious: either an abstract, incoherent identity-in-itself or a relation to otherness that must ultimately be oppositional. This alternative, however, is viable only if noble morality affirms self-identity.

What interests Deleuze about noble morality, however, is that it affirms an immediate differentiation exceeding identity and opposition—that is, a relationship of irreducible heterogeneity, of internal quantitative difference. Deleuze cites two central Nietzschean ideas: that the friend is someone between "I" and "me" who helps me overcome myself, and that going beyond good and evil does not mean going beyond good and bad—which does not imply its identity with the ethic of good and bad (Deleuze 1983, 6, 122). When the noble calls himself good, he affirms not identity, which would require opposition to another, but an ability to transcend limits and overcome himself. This overcoming requires a relation to another, who need not be recognized as the "same" but only as sufficiently strong to offer a challenge. This other must therefore be another noble, but the term "noble" does not establish an identity or sameness. Immediate affirmation thus embodies not a withdrawal into self but rather a pluralism[18] whereby the noble affirms himself, his adversary, and their struggle. Affirmation is immediate not because it lacks relation but because its relation is not a measurement or comparison of oneself to another via a fixed hierarchy of values. Nobility does compare itself to the weak, but as a consequence of its defining goodness as overcoming, not as a precondition of this definition: "no comparison interferes with the principle. It is only a secondary consequence, a negative conclusion that others are evil insofar as they do not affirm, do not act, do not enjoy" (120). This self-differentiation, achieved through a relation of strife (hence difference) with another, is misunderstood as selfishness and egoism by slave morality, which sees only identity and opposition: "the difference between forces seen from the side of reaction becomes the opposition of reactive to active forces" (125).

Acting to pluralize, overcome identity, and dissolve oppositional relations, noble morality acts and affirms the eternal return. The eternal return, Deleuze says, is a return not of identical events in history—that given an infinity of chronological time all events will eventually repeat themselves—but of immediate differentiation: "We misinterpret the expression 'eternal return' if we understand it as 'return of the same.' . . . It is not some one

18. "Nietzsche's philosophy cannot be understood without taking his essential pluralism into account" (Deleuze 1983, 4).

thing which returns but rather returning itself is the one thing which is affirmed of diversity or multiplicity" (Deleuze 1983, 48). The return must therefore be understood as a synthesis constituting the passage of time. It does not connect different moments in chronological order but constitutes the internal structure of the moment itself: "it is the synthetic relation of the moment to itself, as past, present and to come, which absolutely determines its relations with all other moments. The return is not the passion of one moment pushed by others, but the activity of the moment which determined the others in being itself determined through what it affirms" (193; on this point see also Widder 2002, 44–48). This is how the eternal return offers "an explanation of diversity and its reproduction, of difference and its repetition" (Deleuze 1983, 49). What returns, however, is the disequilibrium of forces that relate through an internal quantitative difference that includes both power and resistance. The synthesis of the eternal return answers the dialectical synthesis of identity, providing an alternative to dialectical sense. In place of mediation relating and speaking through all differences, there is immediate disjunction. But these force relations also engender a perspective that denies this sense and reduces it to opposition. By eliminating the abstraction of equality that underpins dialectics, a new philosophy of sense emerges, one that, like Hegel's, denies the indifferences of traditional metaphysics, seeing these indifferences as products of difference and its synthetic operation.

7

Abstract and Concrete Differences: Lacan and Irigaray

AS KOJÈVE (1969, ESP. "In Place of an Introduction") explains, Hegelian desire consists of two levels. Natural desire seeks to negate the otherness of an object in order to possess or consume it. Human desire, however, negates this negation, though it also preserves and sublates both natural desire and the other that natural desire would consume. Human desire desires not to negate some other but to negate itself, becoming an object of desire for another. This level of desire, which expresses itself in the demand for recognition by a worthy other, drives the dialectic in *The Phenomenology of Spirit* to its conclusion in the moral society of reciprocal recognition. Hegel's promise is that such reciprocity can be actualized, securing the identity of each counterpart in an Identity of identity and difference.

In contrast, Lacanian desire exceeds this negativity. It goes beyond the satisfaction of (material) need and the demand for love (from the Other), but it is also generated by their interplay. In the first place, there is "a deviation of man's needs from the fact that he speaks," such that these needs, being articulated to another, "are subjected to demand" and thus "return to him alienated" (Lacan 1977, 286). Demand, in turn, reconfigures every satisfaction of need into proof of love and thereby "annuls (*aufhebt*) the particularity of everything that can be granted," while it "constitutes the Other as already possessing the 'privilege' of satisfying needs, that it [*sic*] is to say, the power of depriving them of that alone by which they are satisfied." But this canceled particularity of need returns as a residue of desire "*beyond* demand" (286), as no particular gift can demonstrate the unconditional love demand requires. Even when the subject is given everything, a lack remains and a nameless "something" is withheld, resulting in an inversion of the subject's relation to the Other. Whereas the subject's demand is directed to the phantasm of an omnipotent Other able to provide satisfaction, and calls for a Law to check the Other's caprice, desire, presenting itself "as autonomous in relation to this mediation of the Law," reverses this status: the subject and its demand are now submitted to the Other's desire, which, establishing the Law and subverting the recognition that would ensure subjectivity,[1] rises "to the power of absolute condition (in which 'absolute' also implies 'detachment')" (311).

> By a reversal that is not simply a negation of the negation, the power of pure loss emerges from the residue of an obliteration. For the unconditional element of demand, desire substitutes the "absolute" condition: this condition unties the knot of that element in the proof of love that is resistant to the satisfaction of a need. Thus desire is neither the appetite for satisfaction, nor the demand for love, but the difference that results from the subtraction of the first from the second, the phenomenon of their splitting (*Spaltung*). (287)

Desire thus introduces an indispensable but unresolvable element into the constitution of the subject, and "there is no firmer root" separating the Hegelian from the Freudian subject than "the modes that distinguish the dialectic from desire" (301). The residue of Lacanian desire is an enigmatic and alienated "something" that "constitutes an *Urverdrängung* (primal repression), an inability, it is supposed, to be articulated in demand" (286). The failure of the dialectical reconciliation of identity through difference thereby indicates something that exceeds identity and dialectical difference.

How this excess is implicated in the constitution of subjectivity indicates a dialectical and an extradialectical form of slippage. A subject initially gets a sense of itself through an image conveyed from without, so that unity always comes paradoxically from a passage through an outside and a relation to others.[2] Nevertheless, the subject only truly comes into being on the

---

1. The Other's desire "does not recognize me, nor does it misrecognize me [. . .] . It calls me into question" (Lacan, quoted in Pluth 2006, 304).

2. Boothby (1991, 41–45) argues that the imaginary alienation of the mirror stage precedes the alienating relation of self to other in Hegel's master-slave dialectic. However, matters are more complicated. On the one hand, in the mirror stage, "the *I* is precipitated in a primordial form, before it is objectified in the dialectic of identification with the other, and before language restores to it, in the universal, its function as subject" (Lacan 1977, 2). On the other hand, the mirror stage is itself grounded in the infant's earlier interest in the human form and human face and so with relations to others that are consolidated in its encounter with the mirror image: "What we have there is a first captation by the image in which the first stage of the dialectic of identifications can be discerned. It is linked to a *Gestalt*

terrain of the Other, through the trauma of the phallic Law of the Father, which must be repressed. Trauma suggests an original unity that has been fractured—the castration complex thus retroactively gives meaning to the mirror stage, even while the latter prepares the way for the former—but this imaginary unity never actually existed. The subject thus comes into being carrying an inexpressible feeling of loss. Moreover, since what is felt to be missing was not part of a real prior unity, it is unrecoverable. The subject's search to restore its fullness thus becomes a desire for an impossible lost object—a desire beyond demand—that has already been forbidden to it by a mysterious authority that takes this object as its own desire. Always incomplete and alienated, the subject seeks precariously to establish itself in relation to existing others, confirming its split nature even while striving to repair itself; yet no actual other can sufficiently fill the void created by the lost object. The subject's foundation being also the foundation of language, the two forms of displacement also appear in the linguistic structure. On the one hand, signifiers suffer metonymic displacements—dialectical slippages whereby the part stands for the whole—that introduce lack and negation into signifying chains by the way their signifieds shift beneath them.[3] On the other hand, the creative moment of language is given by metaphorical substitution, when the signifier crosses the barrier separating it from the signified[4]—a crossing that cannot itself be signified (152)—and replaces a traumatic moment that is not merely an absent referent but an absence never able to become present. Metaphor thus "occurs at the precise point at which sense emerges from non-sense" (158), at the frontier where the unnameable makes itself felt within language. Metaphor founds language through a constitutive exclusion—even if the excluded is not prior to the language that its exclusion establishes—that returns as a specter haunting

---

phenomenon, the child's very early perception of the human form, a form which, as we know, holds the child's interest in the first months of life, and even, in the case of the human face, from the tenth day. But what demonstrates the phenomenon of recognition, which involves subjectivity, are the signs of triumphant jubilation and playful discovery that characterize, from the sixth month, the child's encounter with his image in the mirror" (18).

3. "The metonymic structure . . . indicat[es] that it is the connexion between signifier and signifier that permits the elision in which the signifier installs the lack-of-being in the object relation, using the value of 'reference back' possessed by signification in order to invest it with the desire aimed at the very lack it supports" (Lacan 1977, 164).

4. "The metaphoric structure indicat[es] that it is in the substitution of signifier for signifier that an effect of signification is produced that is creative or poetic, in other words, which is the advent of the signification in question. The sign + between ( ) represents here the crossing of the bar — and the constitutive value of this crossing for the emergence of signification" (Lacan 1977, 164).

the signifying chain.[5] There are thus instabilities internal to language, inso-far as its signifiers refer outside themselves to other signifiers, so that each one "assumes its precise function by being different from the others" (126). But a more fundamental instability arises from language's founding split/substitution, which both relates it to a nameless excess and establishes the mirage of unity and identity by burying this excess.[6]

Lacan thereby presents two sorts of relation:[7] a dialectical relation, which establishes meaning and subjectivity by referring any identity beyond itself and which, if there were nothing more, "is convergent and attains the con-juncture defined as absolute knowledge" (296), and another relation to an enigmatic Other, or "a second degree of otherness" (172), which is the locus of the signifying chain constituting the subject, but which makes any full resolution impossible. While the former operates on the terrain of possible identifications—defining, for example, the dichotomy of masculine and feminine around the being or having of the phallus and thereby outlining positions actual men and women may assume (see 289–91)[8]—the latter, the phallus itself, both underpins this oppositional terrain and precludes any identification consolidating itself firmly. On the one hand, the phallus, as privileged signifier, marks all signifiers with the power to stand over their signifieds: "it can play its role only when veiled, that is to say, as itself a sign of the latency with which any signifiable is struck, when it is raised (*aufgehoben*) to the function of signifier. . . . The phallus is the signifier of this *Aufhebung* itself, which it inaugurates (initiates) by its disappearance" (288). This power relates to a paradoxical unity whereby the phallus signifies its own sense, unlike conventional signifiers whose sense must be signified by other signifiers: "As such it [the phallus] is inexpressible, but its opera-tion is not inexpressible, for it is that which is produced whenever a proper noun is spoken. Its statement equals its signification" (316–17). On the

5. "But if in this profusion the giver has disappeared along with his gift, it is only in order to rise again in what surrounds the figure of speech in which he was annihilated" (Lacan 1977, 157).

6. "But if there still remains something prophetic in Hegel's insistence on the funda-mental identity of the particular and the universal, an insistence that reveals the measure of his genius, it is certainly psychoanalysis that provides it with its paradigm by revealing the structure in which that identity is realized as disjunctive of the subject, and without appeal to any tomorrow" (Lacan 1977, 80).

7. Boothby (2001, 150–63) similarly locates two forms of dismemberment of the imagi-nary ego, one related to the symbolic and the other to the real.

8. The function of the phallus in this (hetero)sexual game is essentially captured by the description of Mike Myers's character, Austin Powers: women want him; men want to be him.

other hand, the phallus is always missing from its place, like the dead Father whose authority it represents: "No doubt the corpse is a signifier, but Moses's tomb is as empty for Freud as that of Christ was for Hegel" (316). An oblivion on "a more primordial level, structurally speaking, than repression" (Lacan 1981, 27) makes the phallus "the signifier for which all the other signifiers represent the subject: that is to say, in the absence of this signifier, all the other signifiers represent nothing, since nothing is represented only *for* something else" (Lacan 1977, 316). Communicating its power via its absence, the phallus is the groundlessness that enables all signifiers to function by installing itself in them as a haunting lack or scission. Meaning becomes possible only through this traumatic split, and everything that *is*, that has sufficient unity to *be*, has this status only by being marked by the phallus as split, traumatized, and castrated. Only to the degree that this trauma is repressed, however, can partially stable identifications arise.

The second-order difference is in no way singular: it is possible to disaggregate several kinds of namelessness, all of which are excluded from discourse because they remain unmarked by the phallic Law. These include the phallus itself, which threatens and splits, but whose unity differs from the split unity of all representable things, giving it a mysterious, transcendent, and divine status; the lost object of enjoyment (*jouissance*), the *objet a*, which the Law prohibits and for which the subject must repress its desire, accepting other enjoyments instead; and the feminine, which is defined through the Law as lacking a phallus, being unmarked by it, or being marked only as not being so marked and which thereby becomes both an object of desire and a mysterious truth. A central role of analysis, for Lacan, is to perform this disaggregation, to separate the imaginary abundance of the object of desire from the symbolic excess that points to transcendence, and thereby "to dissociate the *a* and the O [the symbolic Other], by reducing the former to what belongs to the imaginary and the latter to what belongs to the symbolic" (Lacan 1982, 153–54). The feminine, however, is particularly plagued by an unresolvable dual status, because it resides along the two axes of difference. On the one hand, the feminine is located within language as the opposite or complement to the masculine, but this oppositional identity is always inadequate—hence the mystery of women. On the other hand, the feminine is designated, again within language, as a nameless excess (146). For these reasons, the masculine/feminine dichotomy is only partially designated by the active/passive binary, which serves "to name, to cover, to metaphorize that which remains unfathomable in sexual difference. . . . As such, the masculine/feminine opposition is never attained" (Lacan 1981, 192). At once on the margins of language and seemingly beyond it, the feminine both completes the masculine and subverts the relation between them. This allows the feminine to be closely associated with the divine and creates the mystery of feminine enjoyment, which men clumsily define as "vaginal" or "not clitoral" but which cannot be articulated even by the women who experience it (see Lacan 1982, 145–47). It also underpins Lacan's declaration that "~~The~~ woman" does not exist (see 143–44, 151).

Lacan seemingly goes beyond Freud on the question of feminine desire. Freud denies the possibility of two libidos, insisting that it is appropriate, even if strictly speaking incorrect, to call the one libido masculine. He proceeds through this single libido theory to resolve the mystery of femininity, but in so doing domesticates the very question of sexual difference he says has always plagued men: women are defined through representational or oppositional terms of more or less—more passive, having less developed superegos, and so forth—but if this were *all* that differentiated male and female sexuality, no mystery of femininity would exist in the first place (see Freud 1961d; 1965, 112–35). Lacan, in contrast, maintains that if feminine desire exists, it remains unrepresentable: hence neither women patients nor analysts have ever been able to say what it is.[9] At best it "exists" as that which "is not," for what exists and has the status of a thing can be such only by virtue of language.[10] Any attempt to retrieve this feminine desire, Lacan argues, violates the basic principles of psychoanalysis by positing the feminine as a truth prior to signification, presupposing exactly the sexual difference that requires explanation.

This position, however, rests solely on commitments that indicate a continuing reliance on dialectical understandings of difference. These commitments make opposition or contradiction the only possible meaningful relation and conceive any excess as something that must be consigned to oblivion to make this relation possible. The structures of subjectivity, lan-

9. "What gives some likelihood to what I am arguing, that is, that the woman knows nothing of this *jouissance*, is that ever since we've been begging them—last time I mentioned women analysts—begging them on our knees to try to tell us about it, well, not a word! We have never managed to get anything out of them. So as best we can, we designate this *jouissance*, *vaginal*, and talk about the rear pole of the opening of the uterus and other suchlike idiocies" (Lacan 1982, 146).

10. "For it is still not enough to say that the concept is the thing itself, as any child can demonstrate against the pedant. It is the world of words that creates the world of things—the things originally confused in the *hic et nunc* of the all in the process of coming-into-being—by giving its concrete being to their essence, and its ubiquity to what has always been" (Lacan 1977, 65).

guage, the unconscious, and the kinship exchange that underpins culture are, for Lacan, all linked in this same foundation: "the Oedipus complex . . . may be said, in this connexion, to mark the limits that our discipline assigns to subjectivity: namely, what the subject can know of his unconscious participation in the movement of the complex structures of marriage ties" (Lacan 1977, 66). Yet, while these structures inevitably fail to secure themselves because they are constituted on the terrain of the Other, which both enables and destabilizes any meaningful—that is, dialectical—relations precariously established on it, alternatives cannot be theorized because only these forms of subjectivity, language, and culture are available: "the elementary structures of culture . . . reveal an ordering of possible exchanges which, even if unconscious, is inconceivable outside the permutations authorized by language"; "culture . . . could well be reduced to language, or that which essentially distinguishes human society from natural societies" (148). Thus, while involvement of the Other is sufficient to distinguish the Lacanian subject from both the classical philosophical subject theorized from Descartes to Hegel and the perceptual and linguistic centers represented by the "ego," and the "I,"[11] it is essentially a limit concept for these manifestations of identity and their restricted conceptions of difference: it decenters them, but they continue to define the nature of meaning and sense in a (dialectical?) circle of identification and its failure.[12]

For Irigaray, all this is indicative of a "masculinist" logic and economy of sameness and their narrow conception of difference. The problem, as she sees it, is not that this thinking cannot acknowledge an Otherness that is irreconcilable *within its terms,* but that it nevertheless conceives this irrecon

11. "In short, we call ego that nucleus given to consciousness, but opaque to reflexion, marked by all the ambiguities which, from self-satisfaction to 'bad faith' (*mauvaise foi*), structure the experience of the passions in the human subject; this 'I' who, in order to admit its facticity to existential criticism, opposes its irreducible inertia of pretences and *méconnaissances* to the concrete problematic of the realization of the subject" (Lacan 1977, 15; see also 23, 60, 80, 89–90, 126–28, and 307).

12. Thus, on the one hand, "Freud's discovery was to demonstrate that this verifying process authentically attains the subject only by decentring him from the consciousness-of-self, in the axis of which the Hegelian reconstruction of the phenomenology of mind, maintained it" (Lacan 1977, 80). However, on the other hand, "in the term *subject*—this is why I referred it back to its origin—I am . . . designating . . . the Cartesian subject, who appears at the moment when doubt is recognized as certainty—except that, through my approach, the bases of this subject prove to be wider, but, at the same time much more amenable to the certainty that eludes it. This is what the unconscious is" (Lacan 1981, 126). For some, it is a virtue of Lacan's subject that it is constituted by its own impossibility, because this saves the subject from its "poststructuralist" deconstruction and reduction to a mere subject position. See Stavrakakis (1999, 13–15), and Žižek (1989, 72, 173–75).

cilability *on its own terms* as something that cannot be spoken, understood, or recognized. Unsurprisingly, Lacan's various presentations of unnameable Otherness repeat the moves of classical metaphysics either to elevate it to divine status or to reduce it to abyss, chaos, or materiality. This prompts Irigaray to ask, "Might psychoanalysis, in its greatest logical rigor, be a negative theology? Or rather the negative of theology?" (Irigaray 1985, 89).[13] The result is that psychoanalysis remains unable to examine its historical and genealogical origins.[14]

In the phallic economy that Lacan relates to language, culture, and subjectivity, the feminine is subjected to abstractions that parallel those Marx outlines in commodity exchange: *"what is required of a 'normal' feminine sexuality is oddly evocative of the characteristics of the status of a commodity"* (Irigaray 1985, 187). Commodities are exchanged only through an abstraction that establishes equivalences among qualitatively distinct goods, allowing each good to have a quantitative value in relation to another commodity held as a fixed value. Buyers and sellers in the marketplace reciprocally recognize each other as subjects, but only by a detour through commodities, which are simultaneously treated as objects and endowed with a value transcending the commodities themselves, thereby enslaving the men who exchange them. The commodity thus inhabits two positions, one involving an abstract (but representative) value in a dialectic of exchange and the other involving an enigmatic value that makes the commodity stand for the promise to satisfy an impossible, insatiable desire. These forms of difference and the production of a mysterious and transcendent Other have their structural counterparts in the phallic economy Lacan declares to be indispensable.

> Thus, starting with the simplest relation of equivalence between commodities, starting with the possible exchange of women, the entire enigma of the money form—of the phallic function—is im-

13. Irigaray presents her own feminine divinity, but it is decidedly opposed to the transcendence of the divine over the human. On this point, see Deutscher (1994). On the resonances between Irigaray's sensible transcendental and Deleuze's immanent virtual and between Irigaray and Deleuze more generally, see Lorraine (1999). Of course, Irigaray's objections to Deleuze and Guattari's thesis on becoming-woman should be noted. On this point, see Irigaray (1985, 140–41); see also Braidotti (1994, 168–71), and Grosz (1994, 189–90).

14. "Psychoanalytic theory thus utters the truth about the status of female sexuality, and about the sexual relation. But it stops there. Refusing to interpret the historical determinants of its discourse . . . and in particular what is implied by the up to now exclusively masculine sexualization of the application of its laws, it remains caught up in phallocentrism, which it claims to make into a universal and eternal value" (Irigaray 1985, 102–3).

plied. That is, the appropriation-disappropriation by man, for man, of nature and its productive forces, insofar as a certain mirror now divides and travesties both nature and labor. Man endows the commodities he produces with a narcissism that blurs the seriousness of utility, of use. Desire, as soon as there is exchange, "perverts" need. But that perversion will be attributed to commodities and to their alleged relations. Whereas they can have no relationships except from the perspective of speculating third parties. (177)

If this economy of sameness rests on abstraction, what does it abstract from? Just as Marx held that commodity exchange effaces both the real qualities of goods and the real relations constituting them, Irigaray answers that what is abstracted away is the feminine, but the feminine is a relation, not a substance. The character of this relation and its constitutive role can be understood in terms of mirroring. The philosophy of subjectivity, including the Lacanian variant, Irigaray says, depends on "recourse, explicitly or more often implicitly, to the *flat mirror*" (154). An image passing between two flat mirrors remains the same, and although something is not mirrored, such as the back side of each mirror, this matters little insofar as the movement of the image is concerned: the tains of the mirrors make the mirroring possible, but even granting this point the reflection continues to be one of identity. But everything changes if the mirrors are curved: the result is an anamorphosis, an image spinning out of control, becoming amorphous and liquid.[15] No point relates in a clear one-to-one way with its reflection, yet all points are related nonetheless. Masculine logics, Irigaray maintains, abstract the curvature from the mirrors, reconfiguring the difference that actually structures and contours relations of difference and making it reappear as the excluded remainder of a reflexively constituted identity. The result is a feminine conceived in terms of lack, absence, and failure. By contrast, removing these abstractions brings to light a feminine relation that, structuring relations through decentering, invokes a nearness, but "a 'near' not (re)captured in the spatio-temporal economy of philosophical tradition" (153–54).

While the abstractions of dialectical contradiction indicate an excessive difference, excessiveness is not enough to eliminate this difference's own abstractions. Insofar as difference or Otherness is still understood from a

---

15. See Irigaray (1985, 154–55). Irigaray's idea should be compared with Lacan's reference (1981, 85–89) to anamorphosis and the gaze of the Other, noting, however, that Lacan does not make use of it in his account of the feminine.

perspective that privileges a dialectical construction of subjectivity and sense, and insofar as the limits of dialectical opposition are traced not in order to deny its necessity but to reveal its paradoxical conditions of being, difference is conceived in a way that effects a return of identity and transcendence. Irigaray's move, in this regard, razes these abstractions by weaving second-order difference into another form of meaningful synthesis, one that links differences without drawing them into unity, thereby securing "a place for the feminine within sexual difference" (159). Considered concretely, then, second-order difference is not simply a limit to dialectical synthesis but the structuring principle of a disjunctive synthesis. Giving such a content and role to this difference neither creates an inclusive unity nor posits a pregiven essence, as some have argued.[16] It is rather the move that completes immanence, as required by a philosophy of sense.

16. Judith Butler, for example, attributes to Irigaray a desire not to exit discursive structures but to articulate "an entirely different economy of signification" (1990, 10; see also 9–13, 18–19, and 103) that breaks with identity-based economies. Nevertheless, Butler refuses to pursue Irigaray's alternative signifying economy very far, suggesting that unless it takes the form of a subversive miming of masculine signification that remains within the latter, it reinvokes essentialism (1993, 47–48). More generally, Butler maintains that any rejection of the notion of constitutive exclusion ultimately falls back on the Hegelian pipedream of inclusive unity: "The ideal of transforming all excluded identifications into inclusive features . . . would mark the return to a Hegelian synthesis which has no exterior and that, in appropriating all difference as exemplary features of itself, becomes a figure for imperialism, a figure that installs itself by way of a romantic, insidious, and all-consuming humanism" (1993, 116). Consider, in this regard, her critical readings of Foucault and Deleuze (1987, 186–238) and her critique of Foucault's reading of Herculine Barbin (1990, 93–106).

8

Repetition and the Three Syntheses of Time

HOW DOES DIFFERENCE become the structure of time as such, rather than
an event occurring in time? For Deleuze, it is a matter of repetition. Repetition must not be confused with generality, which belongs to the domain of
law (Deleuze 1994, 2) and presents a "qualitative order of resemblances"
and a "quantitative order of equivalences" (1).[1] Law relies on abstraction and
closed systems (3) to sanction reiterations of identity, as though two events
or two things could ever *really* be the same. Law and generality treat the
differences between repetitions with *indifference,* as secondary or accidental,
but genuine repetition does not allow this and so breaks law and identity.
Repetition, in short, repeats difference. And yet, Deleuze argues, we must
distinguish levels of repetition: "To repeat is to behave in a certain manner,
but in relation to something unique or singular which has no equal or equivalent. And perhaps this repetition at the level of external conduct echoes,
for its own part, a more secret vibration which animates it, a more profound,
internal repetition within the singular" (1). Within the multiplicity of actual
repetitions and their differences, then, is a constitutive repetition of internal
difference, which synthesizes, or "contracts," differences.

Internal difference is an immanent, ontological difference capable of carrying out the requisite synthesis. The error of traditional metaphysical philosophy, Deleuze maintains, is that it never reaches this "difference in
itself" because it confuses it "with a merely conceptual difference" (27),
seeking to accommodate difference to identity rather than elaborating a concept of difference that differs from both identity and identity's conception
of difference. Even the Hegelian alternative, speculative contradiction, aims

1. For Bergson (1998, 223–31), the confusion and conflation of these two orders, which
he calls the generality of genera and of law, reduces vital order and repetition to geometrical
organization.

to secure identity, specifying the identity of a thing by differentiating it from everything it is not while simultaneously mediating this difference. Contradiction, however, is the maximum form of difference, and therefore the only difference able to fill this synthesizing role, "only to the extent that difference is already placed on a path or along a thread laid out by identity" (49–50). Moreover, it achieves its end only abstractly. A more concrete disjunctive synthesis, which maintains the heterogeneity of the relations it brings together, implies a "second-degree difference" (117) or differenciator exceeding the terms of identity, resemblance, and equality.

Deleuze thus holds this differenciator to be an enigmatic difference, whose only "identity" can be as that which differs from itself, never being where it might indicate it is. Deleuzean repetition here converges with Lacanian repetition. For Lacan, repetition is never mere reproduction but mimicry, not passive adaptation to or copying of surroundings but active inscription[2] of the subject into a picture (Lacan 1981, 99) through the assumption of the perspective of the gaze of the Other. Mimicry thus involves disguise, camouflage, and masquerade: "To imitate is no doubt to reproduce an image. But at bottom, it is, for the subject, to be inserted in a function whose exercise grasps it" (100). Ultimately, mimicry works by lure—a sexual lure,[3] playing on desire—and need not accurately mirror reality (see 111–12). It evokes the power of the simulacrum as an appearance pretending to be the truth behind appearances—an appearance that masquerades as an essence by hiding the fact that there is nothing behind the mask (112). Deleuze similarly holds that in the game of repetition the differenciator generates masks, yet there is no essence, no identity beneath the masks, only difference.[4] Masquerade effects the appearance of essence, identity, and resemblance. These, however, are only simulations arising from "a primary difference or a primary system of differences":

> In accordance with Heidegger's ontological intuition, difference
> must be articulation and connection in itself; it must relate different
> to different without any mediation whatsoever by the identical,

2. "The most radical problem of mimicry is to know whether we must attribute it to some formative power of the very organism that shows us its manifestations" (Lacan 1981, 73).

3. "It is no doubt through the mediation of masks that the masculine and the feminine meet in the most acute, most intense way" (Lacan 1981, 107).

4. "Behind the masks, therefore, are further masks, and even the most hidden is still a hiding place, and so on to infinity. The only illusion is that of unmasking something or someone" (Deleuze 1994, 106).

the similar, the analogous or the opposed. There must be a differenciation of difference, an in-itself which is like a *differenciator*, a *Sich-unterscheidende*, by virtue of which the different is gathered all at once rather than represented on condition of a prior resemblance, identity, analogy or opposition. As for these latter instances, since they cease to be conditions, they become no more than effects of the primary difference and its differenciation, overall or surface effects which characterise the distorted world of representation, and express the manner in which the in-itself of difference hides itself by giving rise to that which covers it. (117)

These heterogeneous differences, which cannot be brought into harmony and identity because they pass through a conduit of "difference in itself," are necessarily out of sync with one another. "I" am necessarily out of sync with myself and with my world; consequently I live a mode of repetition, always differing, disguising, and contracting.[5] Freed from its subordination to movement, time can be understood as this contraction, the form of that which, being out of sync with itself, moves or changes in (chronological) time. In the second chapter of *Difference and Repetition*, Deleuze presents this disjointed structure as the last of three syntheses of time, which are associated with Hume, Bergson, and Nietzsche, respectively. Each synthesis displays a form of repetition, the three successive repetitions serving as the foundation, grounding, and ungrounding of time.[6]

The first, Humean, synthesis, which Deleuze calls the passive synthesis of habit, is an empirical repetition/contraction of instants into a line of time, with past and future figuring as dimensions of the present: "Time is constituted only in the originary synthesis which operates on the repetition of instants. This synthesis contracts the successive independent instants into

---

5. Deleuze thus reverses the order of repression and repetition: "I do not repeat because I repress. I repress because I repeat, I forget because I repeat. I repress, because I can live certain things or certain experiences only in the mode of repetition. I am determined to repress whatever would prevent me from living them thus: in particular, the representation which mediates the lived by relating it to the form of a similar or identical object" (Deleuze 1994, 18).

6. Several commentators (for example, Williams 2003; Faulkner 2006) refer to these as three passive syntheses of time, holding that they are passive because they exceed the subject's control. However, Deleuze calls only the first two syntheses passive, precisely because the third synthesis dissolves the subject. The first two syntheses follow the rule that "time is subjective, but in relation to the subjectivity of a passive subject" (Deleuze 1994, 71) and make untenable Kant's final attempt to preserve subjectivity by dividing it into an active transcendental and a passive empirical subject. However, the final synthesis is, properly speaking, neither active nor passive.

one another, thereby constituting the lived, or living, present" (Deleuze 1994, 70). Regardless of whether time is considered the objective measure of movement or the subjective counting of the soul, the self is the site where this synthesis occurs. Moreover, this line really weaves together innumerable microcontractions, which literally go all the way down: there are no instants-in-themselves that are subsequently synthesized; rather, the instants are differentials. Contractions thus repeat and rest upon other contractions. Only with the establishment of this line of time are the active syntheses of memory and intellect possible. The first, passive synthesis is thereby the empirical foundation of time.

This synthesis, however, creates only a static line of time and cannot account for the passage of the present. Thus, while the first passive synthesis provides the empirical foundation for recall and anticipation, it presupposes a second passive synthesis of memory, which provides the transcendental ground of time by enabling the present to pass: "The claim of the present is precisely that it passes. However, it is what causes the present to pass, that to which the present and habit belong, which must be considered the ground of time. It is memory that grounds time" (Deleuze 1994, 79). This is the Bergsonian moment. Bergsonian duration is the reflux of the past into the present, the mark of the past remaining present in such a way that no mechanistic understanding of causality and no linear conception of time can adequately capture the vitality of life (see Bergson 1998, chapter 1). The past remains present, for Bergson, as a virtual temporality folded into the past, present, and future of linear, chronological time. It cannot be composed "after" the present—after the present present passes and another replaces it—because "no present would ever pass were it not past 'at the same time' as it is present; no past would ever be constituted unless it were first constituted 'at the same time' as it was present" (Deleuze 1994, 81). Each present, always already linked to the virtual past, is unique by virtue of this past inhering within it, and the virtual past of one present inheres as a layer of virtual past in subsequent presents. In this way, the virtual past is not merely a past present but the milieu in which we focus on past presents. However, the pure past and its coexistence with the present also constitute the present and its passing, and "for this reason the past, far from being a dimension of time, is the synthesis of all time of which the present and the future are only dimensions" (82). Unlike the first synthesis, the past is not a dimension of the present. Rather, the present is a dimension of the past: it contracts the innumerable layers of virtual past (hence it is a synthesis of *memory*) into actuality, making the present the bursting forth of the past in its creativity and freedom—in Bergson's words, *"time is invention or it is nothing at all"* (Bergson 1998, 341). Like the passive synthesis of habit, the passive synthesis of memory consists of contractions or syntheses that go all the way down, the different levels of the past inhering in and repeating one another.

Nevertheless, duration must give way to another synthesis. Many interpreters closely link Deleuze and Bergson, yet, as already noted, Deleuze moves toward Nietzsche in a way that is not simply a Bergsonian reading of Nietzsche (a position taken by Borradori 2001) or the use of Nietzsche to deepen Bergson (see Boundas 1996, 102).[7] Deleuze's break with Bergson concerns two key considerations of immanence. First, although Bergson introduces difference into time, drawing difference from repetition and demonstrating the impossibility of identical repetitions across time, his insistence on duration's continuity works to safeguard the coherence of the ego—a safeguarding that Bergson explicitly intends[8] and that Deleuze recognizes when he links the formation of the ego to the second synthesis (Deleuze 1994, 108–9). While Bergson's ego is a process, not a substance, duration, in this respect at least, remains a temporality of consciousness[9]—or the ground for such a temporality—whereas Deleuze, in accordance with the requirements of immanence, seeks a time of the "aborted cogito" (110), which demands that a fundamental discontinuity be introduced into time.[10] Second, Deleuze insinuates a continuing transcendence in Bergson, whereby the virtual past, exceeding intellectual conceptualization, remains a mysterious and unrepresentable ground for representation. Deleuze levels this criticism indirectly: after noting that the pure past is

7. Moulard (2002) notes Deleuze's moves away from Bergson in both *Difference and Repetition* and *Cinema 2* but treats it as a radicalization of Bergson without even mentioning Deleuze's turn to Nietzsche.

8. This is clear when Bergson criticizes associationism for trying to recompose consciousness by adding together numerically distinct states, "thus substituting the symbol of the ego for the ego itself" (1910, 226).

9. Bergson makes obvious links between duration and consciousness: "Concrete reality comprises those living, conscious beings enframed in inorganic matter. I say living and conscious, for I believe that the living is conscious by right; it becomes unconscious in fact where consciousness falls asleep, but even in the regions where consciousness is in a state of somnolence, in the vegetable kingdom for example, there is regulated evolution, definite progress, aging; in fact, all the external signs of the duration which characterizes consciousness" (1983, 92).

10. Adorno similarly argues that Bergson remains "within range of immanent subjectivity" (1995, 9). Giorgio Agamben maintains that for Deleuze absolute immanence requires breaking with models of consciousness and takes the form of an ungrounding: "*In Deleuze, the principle of immanence thus functions antithetically to Aristotle's principle of the ground*" (Agamben 2002, 163).

penetrated and lived through reminiscence (84–85), he attacks Platonic reminiscence—which he elsewhere links to Bergsonian memory (Deleuze 1991, 59)—and an equivocation

> already implicit in the second synthesis of time. For the latter, from the height of its pure past, surpassed and dominated the world of representation: it is the ground, the in-itself, noumenon and Form. However, it still remains relative to the representation that it grounds. It elevates the principles of representation—namely, identity, which it treats as an immemorial model, and resemblance, which it treats as a present image: the Same and the Similar. It is irreducible to the present and superior to representation, yet it serves only to render the representation of presents circular or infinite. (Deleuze 1994, 88)

This transcendence of the past suggests that Bergson does not so much overturn a linear, chronological time as simply complicate it, introducing discontinuity into time but also circumscribing discontinuity for the sake of representation and knowledge. The stock Bergson puts in intuition to provide absolute as opposed to relative knowledge, and the respect he accords science's grasp of the material world, should not be overlooked.[11] To the degree that the virtual past and actual present are distinct dimensions of time, they remain unsynthesized, allowing the past to assume a transcendent status. Duration is the concrete time of the concrete ego, but it is nevertheless abstract with respect to an ontology that treats the ego, and identity more generally, as a simulation.

While the coexistence of past and present serves as the transcendental ground of time, resonance around a fracture constitutes time's ungrounding in the third synthesis. The paradox of all three syntheses is that the self constitutes time through repetition, but also resides within time. The self is thereby fractured by time. Kant's critique of Descartes, for Deleuze, demonstrates the consequences. Descartes declares, "I think, therefore I am; I am

---

11. "To metaphysics, then, we assign a limited object, principally spirit, and a special method, mainly intuition. In doing this we make a clear distinction between metaphysics and science. But at the same time we attribute an equal value to both. I believe that they can both touch the bottom of reality. I reject the arguments advanced by philosophers, and accepted by scholars, on the relativity of knowledge and the impossibility of attaining the absolute" (Bergson 1983, 37; see also 38–39, 124, 134, and 161). It should also be remembered that Bergson merely assumes the existence of intuition: "intuition, if it is possible, is a simple act" (162).

a thinking thing." But, Kant replies, the undetermined cannot be determined directly by the determinate—the determinate "I think" implies an undetermined "I am," but this alone cannot show that the "I am" is thereby determined as a "thinking thing." This necessitates, for Kant, the introduction of time as the form of change in which determination takes place: I am a thinking thing *in time*.

> The consequences of this are extreme: my undetermined existence can be determined only *within time* as the existence of a phenomenon, of a passive, receptive phenomenal subject *appearing within time*. As a result, the spontaneity of which I am conscious in the "I think" cannot be understood as the attribute of a substantial and spontaneous being, but only as the affection of a passive self which experiences its own thought—its own intelligence, that by virtue of which it can say *I*—being exercised in it and upon it but not by it. (Deleuze 1994, 86)[12]

Kant, of course, tries to resecure the self's unity in a transcendental subject. While the phenomenal cogito is fractured, its noumenal counterpart's unity is asserted through Kant's restriction of synthesis to the latter. However, once the division between the active and synthesizing transcendental ego and the passive and receptive empirical ego is rejected—and it cannot be sustained, Deleuze maintains, once passive syntheses of habit and memory are acknowledged (87)—what is left is a fractured subject in a time out of joint. The structure of this disjointed time is that of Nietzsche's eternal return.

When presented in linear terms, the eternal return means the return of "difference in itself"—the continual repetition/return of the enigmatic differenciator that always differs from itself. Deleuze sometimes offers this version of the eternal return, especially when opposing it to standard readings that treat it as the idea that, given an infinity of time, identical events will recur endlessly (see esp. 297–301). However, since this differenciator connects heterogeneous differences, neither a linear nor a circular model is adequate to it. Furthermore, as the form of time, the eternal return must be

---

12. Compare with Nietzsche's critique of the cogito: "With regard to the superstitions of logicians, I shall never tire of emphasizing a small terse fact, which these superstitious minds hate to concede—namely, that a thought comes when 'it' wishes, and not when 'I' wish, so that it is a falsification of the facts of the case to say that the subject 'I' is the condition of the predicate 'think'" (1989, §17).

more than a dissymmetry recurring in time. The dissymmetrical structure of time, Deleuze maintains, involves at least two temporal series, which are not successive but coexistent and which come together through a caesura, differenciator, or "dark precursor" circulating through them. This formation has no first term: there may be resemblances and repetitions from one series to the other, but neither series can be an origin that the other simply copies well or badly, because they coexist; nor can the differenciator be considered an origin, because "this would be to assign it a fixed place and an identity repugnant to its whole nature" (105). Rather than an origin of difference, there is only "difference in itself," which, synthesizing differences, repeats and differenciates itself but in doing so repeats difference, not identity: "The eternal return has no other sense but this: the absence of any assignable origin—in other words, the assignation of difference as the origin, which then relates different to different in order to make it (or them) return as such. . . . If difference is the in-itself, then repetition in the eternal return is the for-itself of difference" (125).

When outlining the structure of the unconscious, glimpsed through parapraxes, dreams, and neurotic symptoms, Freud describes a similar resonance among divergent memories, phantasies, and lines of thought. Screen memories arise with the projection of two phantasies onto each other to form a childhood memory (Freud 1962, 301–22); dream work is accomplished through a connection of repressed unconscious material to preconscious residues of the previous day that allows the former to be expressed (Freud 1966, 212); the nodal points organizing dreams are overdetermined by contradictory associations, but also converge on a nodal point about which even Freud sometimes hesitates to speak.[13] The Oedipal nature of this unsayable something is obvious, given that childhood dreams are straightforward and easily interpreted as wish fulfillments, while adult dreams are complicated and obscure (Freud 1952, 15–20). The unconscious processes that condense, displace, and transfer psychic energies obey no rule of temporal order and in fact "have no reference to time at all" (Freud 1957d, 187). Yet Freud maintains that phantasies must be anchored in real childhood experiences (1962, 318–19; also 1963, 205–18; 1966, 369–71) and that hysterical symptoms arise when an auxiliary trauma resonates with an earlier, primary event (Freud and Breuer 1955, 133–34). While adopting this

13. "I could draw closer together the threads in the material revealed by the analysis, and I could then show that they converge upon a single nodal point, but considerations of a personal and not of a scientific nature prevent my doing so in public" (Freud 1952, 10).

general configuration, Deleuze holds that psychoanalysis must dispense with the Freud's pretensions to scientificity and cease being a search for chronological origins: "A decisive moment in psychoanalysis occurred when Freud gave up, in certain respects, the hypothesis of real childhood events, which would have played the part of ultimate disguised terms" (Deleuze 1994, 17). Removing this last element of hidden identity and stability makes the unconscious a domain not where time is absent but where disjointed time is revealed. The Oedipal story, in turn, refers not to a trauma occurring in time but to the traumatic organization of time itself.[14]

In this revised story, the Oedipal trauma may or may not be established by a real childhood event. Its effect, in defining sexual difference in hetero-sexual and genital terms and introducing the castration threat, is at once to separate and join together two orders, one infantile and pregenital and the other adult and genital, each having divergent body images and both real and imaginary objects of desire, memories of the past, and expectations of the future. The expression of this event is the phallus, the signifier of the mysterious paternal Law, which seems to give sense and cohesion to the psyche but is never entirely comprehensible. Because it constitutes the sepa-rate series through a radical break, it cannot be localized within either se-ries, appearing instead at the margins of each. But the phallus does not establish an identity between the series, because it has no identity itself. It is only a marker for a "something is there" or a "something happens" that resides in both series as something vague and unfathomable. It is therefore univocal across the series, but its univocity is that of an enigma.[15] Both series continue within the post-Oedipal unconscious, relating to each other through this "something is there," which acts as a differenciator. It circu-lates a phantasy between consciousness and the unconscious, establishing "the resonance of the two independent and temporally disjointed series" (Deleuze 1990, 226). Infantile and adult series, despite their discontinuity, might seem to succeed each other in time. But this, Deleuze says, overlooks the simultaneously of the series in the unconscious and the fact that no continuous subject exists across the event of trauma—"the two series . . . are not distributed within the same subject" (1994, 124). Moreover, as the traumatic event need not be a real event, it does not mark a chronological

14. Deleuze's use of the Oedipal story is therefore genealogical: rather than searching for ultimate origins, it aims to establish how the structure of the self depends upon a difference and repetition beyond those understood in terms of identity and similarity. On this point see Foucault (1977c, 171), and Lecercle (1989, 89).

15. On Deleuze's use of the medieval thesis of univocity, see Widder (2001).

moment when the self becomes cracked but rather indicates a fundamental fissure: the self is always already cracked.

To be out of sync with myself thus means to be caught up in this way in diverse lines of time referring to different subjectivities. "I" am a multiplicity of subjects living different temporalities within the same, not so unified being. These diverging subjects and times come together by way of their repetition and resonance with one another—the adults one knew or expected to be as child subjects, Deleuze says, resonate in the unconscious with the adult subjects one is among other adults and children (124)—and the series communicate through the crack that is never fully defined for any of these subjectivities but serves as their enigmatic link. This subjective multiplicity can be effaced, leaving the appearance of a single subject living a single line of successive events. The "something happens" may appear as an original or early event, losing its untimeliness by being defined as a "this happened" that was subsequently repressed. One may come to regard one's later loves as repeating a repressed original love for one's mother, forgetting that the different loves do not belong to the same subject or that the love of one's mother may be only part of an adult subjectivity projected onto childhood. These repetitions of apparent similarity or identity, Deleuze argues, result from the perpetual displacement of the differenciator as it flows through the various series and projects an original term in the process of differing from and hiding itself.

> Repetition is constituted only with and through the *disguises* which affect the terms and relations of the real series, but it is so because it depends upon the virtual object as an immanent instance which operates above all by *displacement*. In consequence, we cannot suppose that disguise may be explained by repression. . . . If it [the differenciator] can be "identified" with the phallus, this is only to the extent that the latter, in Lacan's terms, is always missing from its place, from its own identity and from its representation. In short, there is no ultimate term—our loves do not refer back to the mother; it is simply that the mother occupies a certain place in relation to the virtual object [the differenciator] in the series which constitutes our present, a place which is necessarily filled by another character in the series which constitutes the present of another subjectivity. . . . The parental characters are not the ultimate terms of individual subjecthood but the middle terms of an intersubjectivity, forms of communication and disguise from one series

to another for different subjects, to the extent that these forms are determined by the displacement of the virtual object. (105–6)

Only by the establishment of such continuities can there be a linear time of consciousness. This time certainly exists, but it is merely an empirical succession that refers beyond itself, first to a transcendental ground of time and, more profoundly, to time's transcendental ungrounding: "the empirical condition of succession in time gives way in the phantasy to the coexistence of the two series" (125). Linear time is a surface effect of the repetitions and resonances of a temporal multiplicity that generates the optical illusion of identity. When difference becomes the structure of time, this apparent solidity of the ego and this apparent linearity of time give way to the multiplicity of the play of repetition.

Deleuze serially links his three temporal syntheses to the components of Freud's second model of the psyche:[16] the id is formed through passive syntheses of empirical connection that bind excitations to objects; the ego carries out an active synthesis that unifies these bindings, but it is also grounded by a synthesis of virtual memory—the ego is grounded by reminiscence; and the third synthesis both introduces the superego through trauma and ungrounds or dissolves the ego (96–121). Freud correctly links the superego to the death instinct. Nevertheless, Deleuze argues, Freud erroneously restricts himself to a material model of death, consequently conceiving the death instinct as a drive to return to an earlier, inorganic state (111–12). This death is certainly a component of the psyche, but it "is personal, concerning the I or the ego, something which I can confront in a struggle or meet at a limit, or in any case encounter in a present which causes everything to pass" (112). Another impersonal death, however, which Deleuze draws from Blanchot, "refers to the state of free differences when they are no longer subject to the form imposed upon them by an I or an ego, when they assume a shape which excludes *my* own coherence no less than that of any identity whatsoever. There is always a 'one dies' more profound than 'I die'" (113).

Does recognition of this impersonal death have ethical import? It is worth noting the positive ethical role death plays in Hegelian dialectics. For Hegel, the possibility of ethics rests on the reciprocal recognition of self-

16. The links between Deleuze's three syntheses and Freud's metapsychology are analyzed in detail by Faulkner (2006).

conscious moral agency and identity, a process beginning with the master-slave dialectic in the *Phenomenology of Spirit* (Hegel 1977, §§178–96). The encounter with death—a material and personal death—is meant to play a fundamental and formative role. Yet, while the desire to master death's challenge motivates the initial clash of wills, the dialectic can begin only if the encounter is missed. For the victorious master, the struggle necessarily ends too quickly, and his opponent's surrender must leave him unsure of whether his triumph resulted from his own bravery and mastery or his adversary's weakness. The slave also fails to encounter death, as his surrender to fear instead begins his development of self-consciousness and, through the compulsion to work, his experience of himself as an agent. This avoidance demonstrates how Hegel's dialectic rests on a restricted conception of the event of death, one that secures the significance of life and its relation to freedom through opposition. Hegel conceives death as the maximum form of difference compatible with identity: it is the contradiction to be dialectically reconciled with identity, the negation of the ego that enframes and consolidates the ego—and yet the actual event of death remains unrecuperable (on this point, see Derrida 1978). This unrecuperable event of death, no longer compatible with the unification of the ego, resembles more the enigmatic Deleuzean differenciator. Hegelian ethics, premised on the reconciliation of identity, must forego consideration of this difference.

By contrast, the impersonal death precludes this resolution of identity through opposition. Deleuze thereby links it to an ethics that contests the connection usually drawn between ethics and identity. Taking up Nietzsche's thesis that the establishment of identity through opposition—affirming oneself as good by labeling others evil and taking responsibility for oneself by holding others blameworthy—results from profound *ressentiment* towards life, Deleuze calls for affirmation of a cracked self in a world of multiplicity.

> Either ethics makes no sense at all, or this is what it means and has nothing else to say: not to be unworthy of what happens to us. To grasp whatever happens as unjust and unwarranted (it is always someone else's fault) is, on the contrary, what renders our sores repugnant—veritable *ressentiment*, resentment of the event. There is no other ill will. What is really immoral is the use of moral notions like just or unjust, merit or fault. What does it mean then to will the event? Is it to accept war, wounds, and death when they occur? It is highly probable that resignation is only one more figure

of *ressentiment*, since *ressentiment* has many figures. If willing the event is, primarily, to release its eternal truth, like the fire on which it is fed, this will would reach the point at which war is waged against war, the wound would be the living trace and the scar of all wounds, and death turned on itself would be willed against all deaths. We are faced with a volitional intuition and a transmutation. (Deleuze 1990, 149).

This is not solipsism. On the contrary, the solipsist claims ownership of the personal death, ignoring the impersonal death that, opening the self to multiplicity, gives events their ethical import. But if this ethics, in calling for us to be worthy of what happens to us, is not simply a call for resignation in the face of the given,[17] what exactly is affirmed? It is "not exactly what occurs, but something *in* that which occurs, something yet to come which would be consistent with what occurs, in accordance with the laws of an obscure, humorous conformity" (149). The significance of events is thus that they expose the untimely crack within us that puts us out of joint with ourselves: "the war, the financial crash, a certain growing older, the depression, illness, the flight of talent. But all these noisy accidents already have their outright effects; and they would not be sufficient in themselves had they not dug their way down to something of a wholly different nature which, on the contrary, they reveal only at a distance and when it is too late—the silent crack" (154–55). Affirming this crack depersonalizes events—they cease being the wounds of an ego that suffers them at every turn. Whereas the man of *ressentiment* dies innumerable times with each petty wound and minor sore, affirmation expunges *ressentiment* and, through transmutation, turns these events back upon themselves: "It is at this mobile and precise point, where all events gather together in one that transmutation happens: this is the point at which death turns against death; where dying is the negation of death, and the impersonality of dying no longer indicates only the moment when I disappear outside of myself, but rather the moment when death loses itself in itself, and also the figure which the most singular life takes on in order to substitute itself for me" (153).

The form of this affirmation is the eternal return. In one sense, the test of the eternal return, Nietzsche says, is to transform every "it was" into

17. Such resignation, Deleuze says, characterizes the "beautiful soul . . . who sees differences everywhere and appeals to them only as respectable, reconcilable or federative differences, while history continues to be made through bloody contradictions" (1994, 52).

"thus I willed it" and, more profoundly, into "thus I will its eternal return," thereby expressing the transmutation that makes us worthy of what happens to us. But this formulation is inadequate, because the eternal return "is said of a world *without identity,* without resemblance or equality" (Deleuze 1994, 241) and because its affirmation dissolves both the identity and coherence of the ego and the negative oppositions meant to secure it.[18] Or, rather, the formulation is inadequate because the eternal return affirms that identity and its corollary, opposition, are merely simulations, surface effects generated by divergence: "For if eternal return is a circle, then Difference is at the centre and the Same is only on the periphery: it is a constantly decentred, continually tortuous circle which revolves only around the unequal" (55). The ethical transmutation of the eternal return, therefore, is not meant to secure an "I" who wills. On the contrary, the condition of this ethical affirmation is that the "I," the ego, must be taken far less seriously.

18. In this way the ethical choice of the eternal return differs from a similar structure of choice found in Kierkegaard that sees true choice in a sacrifice that ultimately restores the self to itself. See Deleuze (1986, 114–16).

9

Incorporeal Surfaces

AN ONTOLOGY OF SENSE invokes a surface that brings together and coordinates divergent realms and becomings. It thereby opposes traditional appeals to transcendent Ideas or external *telei*, seeking immanent principles instead. As has been seen, however, apparently anti-Platonist philosophies may still carry residues of Platonism. The same is true of ontologies of the surface. The excess of the surface replaces that of transcendent identities, functioning as a constitutive difference. Yet it may still work to retain the continuity and coherence that are as much a part of Platonism as appeals to transcendence.

The ancient Stoic theory of incorporeals is a surface ontology. The term "incorporeal" (*asōmatos*) is little used in previous doctrines and almost never refers to the Ideas (Bréhier 1997, 1–2).[1] The four incorporeals—"sayables," place, void, and time—are not genera of being, which encompass only corporeal existents, but are instead included with being under a larger category of "something" (see Seneca, *Letters* 58.13–15, 27A, 162; Alexander, *On Aristotle's Topics* 301.19–25, 27B, 162; and Sextus Empiricus, *Against the Professors* 10.218, 27D, 162).[2] Although radically distinct from corporeal bodies, incorporeals are nevertheless indispensable for determining the full significance of these bodies. Platonic Ideas, which are "neither somethings nor qualified, but figments of the soul which are quasi-somethings and quasi-qualified" (Stobaeus 1.136.21–137.6, 30A, 179; see also Diogenes Laertius 7.60–61, 30C, 179), cannot fully determine the finite entities they are sup-

---

1. Perhaps the one example in Plato's dialogues is the *Statesman* 286a, which refers to τὰ ἀσώματα as "the existents which are of highest value and chief importance."

2. All references to ancient sources of Stoic philosophy are from Long and Sedley (1987). Citations consist of the reference to the ancient text followed by the chapter and subsection (e.g., 27A) and page numbers of Long and Sedley's text.

posed to constitute (see Bréhier 1997, 3–4). The Aristotelian alternative of defining beings by dividing genera into species is also uncertain. While the Stoics use division in definition (see Diogenes Laertius 7.60–62, 32C, 190–91), they limit it to conceptual or scientific analysis, and most of their definitions do not take this form (see Long and Sedley 1987, 193). Instead, incorporeals establish the ultimate limits and sense of corporeal things.

The foundation of being is material substrate, but all qualifications and dispositions of matter, which are given in substantial terms and directly affect and modify substance, are likewise corporeal. "Rational" and "animal," as qualifications of substrate, are therefore bodies. So too is virtue, an internal disposition of the corporeal soul, and fatherhood, a disposition of a body relative to another body (see Seneca, *Letters* 113.2, 27B, 176; Simplicius, *On Aristotle's Categories* 166.15–29, 27C, 176; and Galen, *On Hippocrates' and Plato's Doctrines* 7.1.12–15, 27E, 176–77). Dispositions change bodies, but, unlike qualitative changes, do not alter their internal power (Simplicius, *On Aristotle's Categories* 166.15–29, 29C, 176). The four genera of being are therefore "substrates, the qualified, the disposed, and the relatively disposed" (66.32–67.2, 27F, 163). Things have their cohesion, continuity, and qualities by virtue of their *hexis* or "binding spirit." It is composed of a portion of corporeal breath or pneuma that completely interpenetrates bodies without either spirit or body losing its individual properties or ability to separate (on the Stoic theory of mixture, see Sambursky 1959, chapters 1–2). The pneuma itself combines the active material principles of fire and air, water and earth being passive principles (see Nemesius 164.15–18, 47D, 282; Galen, *On Bodily Mass* 7.525.9–14, 47F, 282; and Plutarch, *On Common Conceptions* 1085c–d, 47G, 282). While primary matter constantly alternates without ever retaining its identity, corporeal qualities and dispositions, which determine the form and particularity of bodies, both change bodies and allow them to remain the same (see Plutarch, *On Common Conceptions* 1083a–1084a, 28A, 166; Stobaeus 1.177.21–179.17, 28D, 167–68; Simplicius, *On Aristotle's On Soul* 217.36–218.2, 28I, 169). Nevertheless, these corporeal dynamics of alternation, flux, and identity are incomplete.

The place and role of incorporeals appears in the Stoic account of causality. Causation refers specifically and solely to the interaction of bodies, but in a special way. "Causes are not *of* each other, but there are causes *to* each other" (Clement, *Miscellanies* 8.9.30.1–3, 55D, 334). Causes are corporeal and relate a plurality of bodies, but they produce effects that are incorporeal: "The Stoics say that every cause is a body which becomes the cause to a body of something incorporeal. For instance the scalpel, a body, becomes the cause to the flesh, a body, of the incorporeal predicate 'being cut.' And again, the fire, a body, becomes the cause to the wood, a body, of the incorporeal predicate 'being burnt'" (Sextus Empiricus, *Against the Professors* 9.211, 55B, 333). Incorporeal effects are impassive, "not of a nature either to act or to be acted upon" (8.263, 45B, 272), and differ from the new qualities or properties that causes may induce in bodies. Fire's action in relation to wood and the scalpel's in relation to flesh are causes because they refer to the active corporeal forces of these bodies (see Aetius 1.11.5, 55G, 335). They produce impressions of one body on another or new corporeal mixtures with new properties (Bréhier 1997, 11–12), but these are qualifications that remain within the order of bodies. Effects, on the other hand, are attributes, having the form not of substantives but rather happenings—being or becoming cut or burned. As such, attributes are facts or events. They are neither beings nor properties but are nevertheless affirmed of beings.[3] A correspondence exists between corporeal qualities and incorporeal attributes—the quality of heat that characterizes the mixture of fire and wood relates to the event of being or becoming burned—but they cannot be reduced to a single order. They denote distinct but interconnected levels of forceful beings and surface effects arising in the interstices of these beings.[4]

The actions of external bodies affect or impress themselves on the soul (Aetius 4.12.1–5, 39B, 237), creating thoughts, conceptions, and cognitions, which are corporeal dispositions of the mind (Cicero, *Academica* 2.21, 39C, 237; Aetius 4.11.1–4, 39E, 238). These impressions either relate directly to sensory experiences or further modify the mind (Diogenes Laertius 7.49–51, 39A, 236–37). Thought has the power of utterance (7.49, 33D, 196) and within both thought and its enunciations subsist incorporeal "sayables" (*lekta*), which signify states of affairs: "utterances are voiced but it is states of affairs which are said—they, after all, are actually sayables" (7.57, 33A, 195; see also Sextus Empiricus, *Against the Professors* 8.70, 33C, 196). Sayables have nominative and appellative elements, which denote subjects and common qualities, respectively (Diogenes Laertius 7.58, 33M, 198), but

---

3. "Ces résultats de l'action des êtres, que les Stoïciens ont été peut-être les premiers à remarquer sous cette forme, c'est ce que nous appellerions aujourd'hui des faits ou des événements: concept bâtard qui n'est ni celui d'un être, ni d'une de ses propriétés, mais ce qui est dit ou affirmé de l'être" (Bréhier 1997, 12).

4. "En un autre sens pourtant, ils rendent possible une telle conception en séparant radicalement, ce que personne n'avait fait avant eux, deux plans d'être: d'une part, l'être profond et réel, la force; d'autre part, le plan des faits, qui se joue à la surface de l'être, et qui constituent une multiplicité sans lien et sans fin d'êtres incorporels" (Bréhier 1997, 13; see also Deleuze 1990, 4–7, 20–22).

these elements cannot form propositions without verbs, which correspond to the events or incorporeal effects generated by bodies and which attach themselves to nominative cases (Diogenes Laertius 7.58, 33M, 198; see also Scholia on Dionysius Thrax 230.24–28, 33L, 198). Together these sayables mediate the relations between utterances and the external objects and events to which their meanings refer (Sextus Empiricus, *Against the Professors* 8.11–12, 33B, 195–96) and between corporeal thought and things (Clement, *Miscellanies* 8.9.26.5, 33O, 198). Like incorporeal effects, propositions are impassive—they are affirmed of bodies but belong to another order. The elements of the proposition "Cato is walking" designate a body and express an action, but the proposition is entirely incorporeal, its designating subject being an abstraction of a body that cannot really be distinguished from its activity (see Seneca, *Letters* 117.13, 33E, 196; also Long 1986, 136–37). As propositions, sayables are determinable as true or false. However, while incorporeal propositions can be true, truth itself is corporeal, referring to the disposition of the wise man's soul (Sextus Empiricus, *Outlines of Pyrrhonism* 2.81–3, 33P, 198–99).

Incorporeals may seem to have secondary status insofar as they arise from the interactions of bodies, but their subsistence in the intersections of corporeal substances gives them a constitutive role. This is the crux of Bréhier's thesis on the Stoics, which inspires Deleuze's analysis and use of Stoic incorporeals (Bréhier 1997; Deleuze 1990).[5] The essence of sayables, Bréhier contends, is the verb or "incomplete sayable," which corresponds to the surface effects of bodies and which can stand independently or be attached to a subject in order to underpin the meanings of propositions (1997, 22–23). Moreover, bodies can interact only because of the incorporeals of place and time.[6] Rejecting the Aristotelian conception of a body's place as the surface of whatever body contains it—water's proper place, for example, is the inside surface of the jug containing it (see Aristotle 1934–57, 210b–212a)—the Stoics argue, based on the potentially infinite divisibility that establishes continuity, that the surface extremities of bodies are neither wholes nor parts but rather incorporeals, making it impossible for two bod-

5. Boeri (2001) develops a similar thesis independent of Bréhier and Deleuze.

6. Void's status is distinct, Bréhier notes, because it is not an effect, event, or attribute of bodies, nor is it a surface of the activity of bodies. Moreover, the idea of infinite void surrounding the cosmos implies the relativity and incompleteness of the cosmos, which is contrary to the idea of the world as a complete and living entity. The Stoics were thus led to maintain that no relation exists between the cosmos and void, although they had difficulty holding to this idea consistently. See Bréhier (1997, 44–53).

ies to touch.[7] Since bodies cannot be separated by void, which the Stoics hold cannot exist within the world (Galen, *On Incorporeal Qualities* 19.464.10–14, 49E, 295), it follows that they must interpenetrate, with the consequence that two bodies can occupy the same place—a self-evident impossibility for all other ancient doctrines, but one that enables the mixtures and immanent tensions necessary for the Stoics' surface effects (Bréhier 1997, 40).[8] Essentially, Bréhier argues, the Stoics redefine place as the site of such interpenetration (52–53), making it a product of bodies (a body occupies what would otherwise be empty void, giving place its determinate dimensions [Stobaeus 1.161.8–26, 49A, 294; Sextus Empiricus, *Against the Professors* 10.3–4, 49B, 294]), but also a presupposition for the actions of bodies in which it subsists. Time, Bréhier maintains, is also more than an effect of bodies. Although Zeno defines time as a dimension or interval of motion, Chrysippus calls it a dimension or interval specifically of the world's motion (see Simplicius, *On Aristotle's Categories* 350.15–16, 51A, 304; Stobaeus 1.106.5–23, 51B, 304). Bréhier takes the latter definition to signify a time that is the structure, not the measure, of the world's movement and so of movement and change in general (1997, 55). This temporal structure too is characterized by interpenetration, with past and future subsisting and the present being a mixture of past and future, not an independent, indivisible moment.[9] As an incorporeal, time applies only to events and verbs: the present is the time in which an event is being accomplished and the past is the time when it has finished being accomplished (58). Time is not a cause

7. "Their favourite objection to the champions of partless magnitudes is that there is [i.e. on such a view] contact neither of wholes with wholes nor of parts with parts; for the former produces not contact but blending, while the latter is impossible because partless magnitudes do not have parts. How then do they themselves avoid this trap, seeing that they allow no last or first part? Why, because they say that bodies touch each other by means of a limit, not by means of parts. But the limit is not a body. So body will touch body with something incorporeal, and again will *not* touch, since something incorporeal is in between. But if it will touch, the body will both act and be acted upon by something incorporeal" (Plutarch, *On Common Conceptions* 1078E–1080E, 50C, 299). Nolan (2006) argues that the Stoics affirm an actually infinite division of bodies, places, and time, not into indivisibles but into proper parts that are always further divisible. His conclusions about Stoic conceptions of place and mixture, however, largely parallel Bréhier's.

8. For the Stoics' acceptance that two bodies can simultaneously occupy the same place, see Themistius, *On Aristotle's Physics* 104.9–19, 48F, 292; also Long (1986, 159–60); and Sambursky (1959, 11–17).

9. "It is contrary to the [common] conception to hold that future and past time exist while present time does not, but that recently and the other day subsist while now is nothing at all. Yet this is the result for the Stoics, who do not admit a minimal time or wish the now to be partless but claim that whatever one thinks one has grasped and is considering as present is in part future and in part past" (Plutarch, *On Common Conceptions* 1081C–1082a, 51C, 304).

and does not determine beings or events. It is rather an empty form ("une forme vide" [59]) that contours beings and their surface events, giving them their sense.

Deleuze (1990, 7–8) maintains that the Stoic inversion of Platonism recovers the unlimited chaotic becoming that Plato had sought to bury beneath the supremacy of Ideas. Stoic incorporeal events, linked to verbs, have the character of "infinitives . . . the becoming which divides itself infinitely in past and future and always eludes the present" (5). As far as the Stoics themselves are concerned, these claims are hardly convincing. There is no becoming in two directions at once, as described in *Parmenides*, nor could the Stoics' commitments allow it. Their blended mixtures certainly bring together heterogeneous substances, but any dissonance in these mixtures is subordinate to their unity. Thus, in the case of the most important mixture, "the whole of substance is unified by a breath which pervades it all, and by which the universe is sustained and stabilized and made interactive with itself" (Alexander, *On Mixture* 216.14–218.6, 48C, 290). On the level of bodies, the causal series is submitted to destiny and the eternal recurrence of the same, so that "there will be nothing strange in comparison with what occurred previously, but everything will be just the same and indiscernible down to the smallest details" (Nemesius 309.5–311.2, 52C, 309). Although the Stoics deny that destiny implies necessity, they base this on assertions that possibilities exist that neither are nor will be true (Epictetus, *Discourses* 2.19.1–5, 38A, 230) and that the complex of causes includes human will and character (Cicero, *On Fate* 39–43, 62C, 387), allowing moral culpability to be assigned to actions.[10] These views neither introduce randomness into the causal order nor affirm an Epicurean-like thesis of many worlds that would realize these possibilities. On the surface level, facts may be connected in diverse ways, limited only by the rules of logic that determine their validity (Bréhier 1997, 27). Nevertheless, the Stoics never doubt the possibility of the wise man's infallible knowledge, based on a correspondence of external bodies and events to mental dispositions and sayables. Since rational concepts arise from the traces of real beings impressed on the mind, the two sides can correspond; because the surface events of real bodies and the logical incorporeals embedded in language and thought are neither bodies nor qualities, they can completely coincide (see Bréhier 1997, 18–19, 21–22).

Matters are altogether different, however, if the incorporeal surface is

---

10. On these points see Mates (1953, 36–41), Rist (1969, 112–32), and Sambursky (1959, 57–65, 71–80).

reconfigured as a differenciator, which establishes a dissymmetry and dis-synchrony immanent to the differences it brings together. Mixtures of mate-rial bodies can then have the character of what Deleuze calls in his later writings "rhizomes" and "multiplicities" but in his earlier writings "simula-cra"—"systems which relate different to different by means of difference" (1994, 126) and where "at least two divergent series are internalized . . . neither can be assigned as the original, neither as the copy" (1990, 262). These are the material forms that Plato denigrated as *mere* simulacra or deceptive illusions, locating truth in unchanging Forms. Against this, De-leuze holds the simulacrum to embody positive principles of repetition and difference. The discontinuities of a world of simulacra obliterate any corre-spondence between thought and bodies related through incorporeal say-ables and events, but the challenge this poses to knowledge differs from the traditional problems the Stoics address. It is no longer a question of the wise man's ability to separate true impressions (*phantasia*) from false figments (*phantasma*) (see Diogenes Laertius 7.49–51, 39A, 236–37), and particularly from the case of "a totally indiscernible [but false] impression" (Sextus Em-piricus, *Against the Professors* 7.247–52, 40E, 243), such as that of identical twins. While the Stoics maintain that "no hair or grain of sand is in all respects of the same character as another hair or grain" (Cicero, *Academica* 2.83–5, 40J, 246), this merely asserts the existence of ultimate conceptual differences between seemingly identical things. Their opponents, con-versely, maintain that any differences between indiscernibles remain non-conceptual and indifferent, never affecting their fundamental identity. The simulacrum's differenciator, however, is an order of difference that exceeds the metaphysical dichotomy of truth and falsity. Its effects, therefore, "might be called 'phantasms,' independently of the Stoic terminology" (De-leuze 1990, 7–8). The simulacrum's dynamics, for Deleuze, produce a phantasm that never corresponds to it because the phantasm cannot copy it either well or badly. This phantasm is the surface of sense, but it also gener-ates illusions of identity, continuity, and resemblance: "The same and the similar no longer have an essence except as *simulated,* that is as expressing the functioning of the simulacrum" (262). The reduction of identity to the status traditionally assigned to simulacra is what makes this logic of sense a genuine inversion of Platonism.

While it may not be the case that "the Stoics are the first to reverse Platonism and to bring about a radical inversion" (Deleuze 1990, 7), a route can be taken from the Stoic theory of incorporeal surfaces to an anti-Plato-nist ontology of sense. This path requires, against the continuing Platonisms in Stoic epistemology, the affirmation of a disjunctive synthesis of differences and a fractured structure of time, which together allow sense to emerge from nonsense. In this ontology, the generation of surface sense is accompanied by illusions of identity, which metaphysical philosophy has always considered the sense of being but which has always remained abstract and inadequate to the task. Exceeding the sense given by metaphysics and identity, however, is another sense structured by concrete difference, in which identity is no more than a superficial effect.

10

The Logic of (Non)Sense

CONTRADICTION—the notion that X is at once both itself and not-X—is the nonsense that creates dialectical sense. It provides the mediating surface that stitches together thought and thing, concept and object, what is said and that of which it is said. Deleuze counterpoises difference to opposition, immediate overcoming to mediating contradiction, and disjunction to the Hegelian Identity of identity and difference. Disjunction is now the nonsense that constitutes sense, and it too connects thought and being, but in a way that prevents any simple correspondence between them. The sign now assumes a new role: it is neither the simple reconciliation of thought and reality nor the human site where the Absolute speaks but a paradoxical element causing the heterogeneous domains brought together by the surface to resonate. By locating the inadequacies of Hegel's philosophy of immanence and the pushing internal difference beyond abstract contradiction, Deleuze is able to offer an alternative ontology of sense.

Depth, height, and surface are the three dimensions at play in this new ontology. None can be reduced to the others. Height and depth refer to a hidden Platonic dualism found beneath the more obvious dualism of Idea and copy. An aspect of bodies must escape the action of Ideas or else there would be nothing but Ideas. This aspect neither conforms to nor copies the Idea but rather defines a simulacrum that is neither Idea nor copy. The simulacrum implies a pure and paradoxical becoming that escapes the present, a becoming in two directions at once that characterizes the material and sensible (Deleuze 1990, 2). The surface, in turn, is produced by material bodies and their excessive becomings, but it is irreducible to the material or physical. Here Deleuze adapts the Stoic cleavage between cause and effect discussed earlier. Causality applies to bodies and their interactions, yet, while bodies are causes in relation to each other, their effects are not bodies but events, which form a phantasm that differs from the simulacrum of bodies. Bodies may combine, fall apart, or break one another, but the meaning or sense of these interactions is, following the Stoics, something in excess of bodies. Such excessive events occur on the surface, and while they refer to the bodies that cause them they also relate to each other through a kind of quasi-causality. Events are therefore always submitted to a double relation: the sense they express derives not only from the bodies and interactions that produce them but also from their relation to other meanings or meaningful events. The surface therefore retains an independence from the depths, as the surface tension of water gives it an independence from the depths it covers. Events remain autonomous only insofar as they relate to one another distinctly from their relation to bodies.

Neither body nor Idea, sense is the fourth dimension of propositions, relating concept and thing at and through the surface. Propositions relate directly to things by denoting states of affairs, which may be true or false, possible or impossible; they relate to subjects by manifesting the intentions of a speaking "I"; and they relate to concepts by signifying universal predicates. Each of these three aspects depends on the others: denotation refers to manifestation, since words must express the intentions of a speaker before they can be used to designate states of affairs; yet both manifestation and denotation function only with the constancy of signified concepts; but even though a series of propositions may move from one implied concept to another, it must ultimately affirm a state of affairs as true or false, making signification dependent on denotation (Deleuze 1990, 12–16). Sense, however, cannot be subsumed under these relations: a proposition signifying an incoherent concept (for example, "square-circle") still has sense, and propositions denoting opposite states of affairs (such as "God is" and "God is not") may nonetheless have the same sense. Exceeding the universal and particular of signification, the subject of manifestation and the object of denotation, sense must be a difference that escapes these orders of identity. In this way, Deleuze maintains, sense is a singular event that is best expressed through the infinitive form of verbs. While qualitative states such as "green" and "not green" denote opposite states of affairs ("the tree is green" and "the tree is not green"), the verb "to green" expresses a becoming found in both states of affairs that gives the propositions and the states to which they refer their sense. This sense is also nonsense, expressing a difference whose only "identity" can be as that which differs from itself—a Deleuzean differenciator or a Nietzschean internal quantitative difference. This difference is found only in language and so remains within propositions, but it is attributed to states of affairs without representing these states (21–22). Through the event of sense, the surface of bodies (incorporeal events) and the surface of thought (propositions) meet and interact.

As a field of events, sense not only retains its independence and indifference from the proposition's representational relations to thoughts and things, it also constitutes these relations. Signification must be organized into a hierarchy of concepts, the more general classes encompassing the more specific. But, as is well known, this ordering refers to the paradox of a set of all sets that must include itself as a member, making it at once a highest identity and a member dividing the identity that it presupposes (Deleuze 1990, 68–69). Signification thus invokes a paradoxical sign that, like Lacan's phallus, expresses its own sense—as opposed to conventional signs whose expressed sense must be designated by other signs (28–31). Moreover, signification, denotation, and manifestation all presuppose individual identities, but the individual is a product of a synthesis and, as dialectical synthesis remains abstract, this must be a synthesis of disjunction. On the one hand, the individual is distinguished by the verb events it actualizes. In the Nietzschean conception of forces, this is the role of the will to power, which arises from and determines the convergence of quantitatively different forces. On the other hand, the full identity of an individual is established only with reference to relations extending across divergent worlds: Adam, for example, sins in this world but not in another, incompossible world; Caesar crosses the Rubicon in this world, but in another he is constituted and helps constitute a world in which he does not cross. Only after the individual is established as a nexus of differences can it be predicated of categories, becoming an object of denotation or a subject of manifestation. But because the individual is a product of disjunctive events beyond the order of identity, these predications remain partial and incomplete (109–17). The sense of any proposition thereby continues to refer back to an excessive becoming.

Although sense is a disjunctive synthesis, it differs from the disjunction in the depths of bodies. It differs also from the dimension of heights and Ideas, which invokes identity rather than disjunction. Schizophrenic language collapses words directly into bodies, and the nonsense of schizophrenia destroys any possibility of sense because it cannot express the incorporeality of meaningful events (82–93). The Platonist appeal to transcendent Ideas, on the other hand, remains bulky and abstract.[1] The surface

---

1. Plato's use of irony to appeal to "hypostatized significations," Deleuze says, is countered by humor and monstrous examples, such as Diogenes the Cynic's bringing forth a

subsists only as a persistent noncorrespondence between concept and thing, and sense therefore functions only to the degree that the divergent realms it brings together communicate while retaining their difference. These realms are in fact two series or orders—of word and thing, denotation and expression, what is seen and what is said.[2] Deleuze here speaks of three syntheses, analogous to the three syntheses of time, that together form the field of sense: connective, conjunctive, and disjunctive. The first constitutes a single series by joining heterogeneous elements, the place held by each element determined by its not being another element and not being confused with the others. An example is the different sounds that form words. The second links separate series together, as when the sound patterns of words are associated with corresponding ideas. Finally, the third synthesis causes the series to communicate and the surface connecting them to resonate (see 47, 174–76, 231–32). This disjunctive synthesis invokes an element that circulates through both series while being unidentifiable by either one: "This element belongs to no series; or rather, it belongs to both series at once and never ceases to circulate throughout them. It has therefore the property of always being displaced in relation to itself, of 'being absent from its own place,' its own identity, its own resemblance, and its own equilibrium" (51). The element is the paradoxical or nonsensical sign that underpins denotation, signification, and manifestation. Surface events, differing from themselves and being expressed by the infinitive form of verbs, refer to this pure sign or univocal Event—the Event of eternal return—as a repetition of difference that all events embody (52–65).

An unlikely source of elucidation here is Saussure's linguistic theory—unlikely because Saussure is often thought to hold that language contains only negative differences. Deleuze too holds this reading.[3] But Saussure's thinking is more subtle. He proposes two connective series, one of sounds and one of concepts—signal and significant or signifier and signified. In each series, concepts and sounds find their places through negative relationships to other concepts and sounds. But this negativity holds only for signifiers and signifieds taken in abstraction. Once considered in terms of

---

plucked fowl to answer Plato's definition of man as a featherless biped. See Deleuze (1990, 134–35).

2. Deleuze (1988, 47–69) analyzes Foucault's genealogical method on the basis of this last pair.

3. "Why does Saussure, at the very moment when he discovers that 'in language there are only differences,' add that these differences are 'without positive terms' and 'eternally negative'?" (Deleuze 1994, 204). See also Hawkes (1977, 19–28) for another example of this error. On Deleuze's relation to linguistics, see Lecercle (2002, esp. chapter 2).

meaningful linkages of sound and concept—that is, as signs—another relation appears.

> Everything we have said so far comes down to this. *In the language
> itself, there are only differences.* Even more important than that is the
> fact that, although in general a difference presupposes positive
> terms between which the difference holds, in a language there are
> only differences, *and no positive terms.* . . . But to say that in a lan-
> guage everything is negative holds only for signification and signal
> considered separately. The moment we consider the sign as a
> whole, we encounter something which is positive in its own do-
> main. (Saussure 1986, 166)[4]

Saussure calls this positive relation among signs opposition, as distinct
from (negative) difference.[5] It is not a relation between units in themselves:
"the language itself is a form, not a substance" (169). When properly considering the realm of language, which Saussure compares to a sheet of paper
holding together thought and sound,[6] negative difference cannot account
for meaning and sense. Signs are fully determined only through various
chains of signification that they bring together. Syntagmatic relations are
those a sign has with others in a proposition. Here, "any unit acquires its
value simply in opposition to what precedes, or to what follows, or to both"
(171). Associative relations are those a sign has with signs not in the proposition, such as those sharing its prefix or root, or overlapping in meaning, but
also with signs whose meanings contrast with it. "Quadruplex," for example, links to one chain of signs that includes "quadrupes," "quadrifons,"
and "quadraginta," and to another that includes "simplex," "triplex," and
"centuplex" (178). The sign is thus a nexus where divergent chains of mean-

4. All page numbers for Saussure (1986) refer to the page numbers of the original text
located in the margins, not to the page numbers of the translation itself.

5. "The moment we compare one sign with another as positive combinations, the term
difference should be dropped. It is no longer appropriate. It is a term which is suitable only
for comparisons between sound patterns (e.g. *père* vs. *mère*), or between ideas (e.g. 'father'
vs. 'mother'). Two signs, each comprising a signification and a signal, are not different from
each other, but only distinct. They are simply in *opposition* to each other" (Saussure 1986,
167).

6. "A language might also be compared to a sheet of paper. Thought is one side of the
sheet and sound the reverse side. Just as it is impossible to take a pair of scissors and cut
one side of paper without at the same time cutting the other, so it is impossible in a language
to isolate sound from thought, or thought from sound. To separate the two for theoretical
purposes takes us into either pure psychology or pure phonetics, not linguistics" (Saussure
1986, 157).

ing converge, and these relations give the sign a value independent of its represented meaning or the materiality of the sign's pronunciation (see 150–54).

There is no resonance yet within this structure, nor could there be given Saussure's theoretical commitments. His absolute division between the synchronic and the diachronic reduces linguistic change to historical changes between static, synchronic states. Such transformation is consequently understood in terms of individual modifications that over time mutate the entire synchronic structure.[7] What is missing from Saussure's analysis is the sort of nonhistorical becoming found in Hegel and Deleuze—a becoming within the present moment and internal to the sign. Even while remaining within Saussure's terms of thought, such a becoming might be glimpsed in the repetition of signs, which is not a linguistic potential but must be actual: a sign used only once could not be a sign, so any sign must always already have been repeated. As a general law of signs, this feature is necessarily synchronic (see Saussure 1986, 131). But while dialectical repetition would retain identity through the differences traversed by the sign, Deleuzean repetition would return a difference. What, then, could turn a Saussurean sign into a Deleuzean sign, which "flashes across the intervals when a communication takes place between disparates" (Deleuze 1994, 20)?[8] If "signs remain deprived of sense as long as they do not enter into the surface organization which assures the resonance of two series" (Deleuze 1990, 104), this organization still requires the nonsense sign that differs from itself and is unresolvable within the terms of concept, thing, and the surface connecting them. Indeed, the paradoxical sign is not one surface sign among others, but is rather a crack of the surface. The time of this crack is the eternal return, through which difference and divergence return.

The structure of sense, for Deleuze, requires this crack or empty square, which circulates through the heterogeneous but mutually imbricated layers brought together by the surface. But this emptiness is not a simple absence of sense or a lacuna to be filled. Such ideas impose the same false alternative as metaphysics and dialectics, between structured sense and indifferent chaos: "*either* an undifferentiated ground, a groundlessness, formless non-

---

7. This potentiality is used to explain analogical changes in language (Saussure 1986, 221–30). Saussure admits, however, that even after diachronic changes are accounted for, "there still remains a residue" (196) that creates difficulties in maintaining the synchronic/ diachronic division.

8. On the sign as the object of a "fundamental encounter," see Smith (1996, 30–33).

being, or an abyss without differences and without properties, *or* a su-
premely individuated Being and an intensely personalized Form. Without
this Being or this Form, you will have only chaos" (1990, 106). Surface
sense and surface nonsense are not opposites. Rather, both oppose the absence of sense: "Nonsense is that which has no sense, and that which, as
such and as it enacts the donation of sense, is opposed to the absence of
sense. This is what we must understand by '*nonsense*'" (71). And the reason
for this is that any simple opposition between sense and nonsense is an
abstraction that fails to reach difference. Sense and nonsense are not opposed but disjoined, which is not surprising given that disjunction is a synthesis of difference that exceeds contradiction. Here it becomes clear how
Deleuze both completes and breaks with Hegel: the completion of a philosophy of immanence must move beyond the dialectics of identity and opposition and so must move from a nonsense of contradiction that reconciles
sense and its opposite to a nonsense of difference that constitutes sense in
terms of divergence. This is the direction in which Deleuze takes thought
after dialectics.

11

Regularities of Dispersion

THE ARCHAEOLOGY OF KNOWLEDGE aims to outline the discursive formations that enable the emergence of subjects and objects and that condition the production of knowledge domains. In other words, it seeks to analyze the structures and processes that establish a surface of sense. To achieve this end, Foucault says, "we must rid ourselves of a whole mass of notions, each of which, in its own way, diversifies the theme of continuity" (Foucault 1989a, 21). Negative unities such as "tradition," "author," "oeuvre," "book," "science," and "literature" not only impose themselves from the outside onto discourse but also erect bulky abstractions. Few specificities can be learned about a discourse from the fact that it is signed by one person (23–24); the book always refers to other books, so that its "frontiers . . . are never clear-cut" (23); objects lack the consistency and continuity over time necessary for a science to be defined by its reference to these objects, and the same can be said of the style, concepts, and themes of any science (31–37). It is not that these "ready-made syntheses . . . that we normally accept before any examination" (22) are worthless, but rather that their fuzziness and uncertainty refer them to a different level that constitutes their conditions of emergence: "My intention was not to deny all value to these unities or to try to forbid their use; it was to show that they required, in order to be defined exactly, a theoretical elaboration" (71).

Upon dispensing with these abstract unities, the dispersion of discursive formations can appear. This dispersion is not a simple scattering of elements in an open or empty space; rather, the term here refers to other meanings, found in both English and French, that emphasize disposition and mixture: the dispersion of materials in a building, or, in chemistry, the dispersion or suspension of one substance in another, which constitutes an emulsion or aerosol.[1] Such mixtures, like the Stoics', are not homogeneous; rather, one substance is dispersed in the other while retaining its own properties. Dispersion thereby indicates a difference within the convergence of heterogeneous materials. It is a synthesis of differences but, unlike the commonly accepted syntheses, it does not collapse differences into unity but rather forms the intersection where unities can appear, or at least where the elements can appear that the unities of tradition, science, literature, etc., gather together and organize. It is a synthesis containing discontinuity.

Thus, if we ask what makes a discourse this particular discourse, we may first of all say that it is through a not very well constructed union of elements that separates the discourse from what it is not. But if we ask what allows the elements united under the name of this discourse to appear in the first place, we find that they always arise where divergent discourses intersect. A discursive formation links together heterogeneous discursive zones, from which emerge subjects and objects that fit neither zone completely, these same subjects and objects arising at different intersections. The "surfaces of emergence" of the objects of nineteenth-century psychopathology, for example, are found within the family, workplace, and religious community, while the expert subjects of this new discipline find their legitimacy as "authorities of delimitation" through a convergence of medicine and law that maps the clinical distinction (the clinic itself being formed in the intersection of various medical, political, and other discourses [Foucault 1989a, 50–54]) of health/sickness onto the legal distinction of citizen/criminal, and then brings this to bear on the family, workplace, and religious environments (40–44).[2] Needless to say, the relations between these heterogeneous domains are strife-ridden, and the status of the discipline of psychopathology that emerges from them remains uncertain. A discourse's entities thus appear through what amount to badly tied knots between other discourses, which are themselves formed through badly tied knots—except

1. In physical chemistry, a dispersion is "a type of intimate mixture in which one substance is present in a large number of separate small regions distributed throughout another, continuous, substance; examples are emulsions (one liquid in another) and aerosols (a solid or a liquid in a gas); also, the state of being so distributed" (OED 1989, vol. 4). In French: "État d'une solution colloïdale, en suspension dans un mileu où elle est insoluble" (Robert 1993).

2. Ironically, as Foucault demonstrates with the asylum physician, the authority of these experts involves no actual medical expertise but rests instead on the way they bring together various discursive domains: "The physician could exercise his absolute authority in the world of the asylum only insofar as, from the beginning, he was Father and Judge, Family and Law—his medical practice being for a long time no more than a complement to the old rites of Order, Authority, and Punishment" (1989c, 272).

that it would be wrong to call them "bad" because the knots cannot be tied any other way. There are parallels here with Althusserian overdetermination and condensation (Althusser 1996, esp. chapters 3, 6), but the convergence of these heterogeneous domains, rather than being a causal relation, is one that conditions the possibility of the appearance of subjects, objects, and knowledges. This is why discursive relations are neither primary relations, which establish chains of dependency among real institutional and social forces existing outside of and prior to a discourse taking place, nor secondary relations, which self-reflexively organize objects within discourse (Foucault 1989a, 45–46). They are neither internal nor exterior to discourse but instead "are, in a sense, at the limit of discourse: they offer it objects of which it can speak, or rather . . . they determine the group of relations that discourse must establish in order to speak of this or that object" (46). These relations are therefore inseparable from discursive practices—indeed, discourse itself *is* a practice (46). Within a discursive formation, incompatible or contradictory objects and concepts can coexist, though within the actual discourses emerging from these formations, the law of noncontradiction may apply (62, 65–66).

Two points must be noted. First, the heterogeneous domains such as law and family are themselves loose and porous unities whose subjects and objects are formed in the interstices that link together divergent discourses. Second, the unity of a discursive formation is defined by the regularity of its dispersion—that is, the regularity of the knots it ties—but this regularity must be understood in a very specific way. As the heterogeneous formations are themselves in flux, the regularity is not a static bond: "A discursive formation, then, does not play the role of a figure that arrests time and freezes it for decades or centuries; it determines a regularity proper to temporal processes; it presents the principle of articulation between a series of discursive events and other series of events, transformations, mutations, and processes. It is not an atemporal form, but a schema of correspondence between several temporal series" (Foucault 1989a, 74). As the linkage itself is in motion, the regularity must be seen as a consistency arising from the continuing synthesis of moving series. Like the pattern of bubbles and surface folds that form where two water currents meet, a discursive formation carries with it never fully stable shapes or images—its subjects, its objects, etc.—that persist over time, arising from the interaction of converging flows. Yet the principle that determines the regularity of a discourse does not reside in these visible stabilities. It is not in the subjects or objects themselves, nor in the stability of knowledge, as a discursive formation can persist even while its objects mutate, new objects are found, or new knowledges are discovered or forgotten ones are rediscovered. Instead, Foucault says, the coherence of the formation lies in the regularity of statements.

A statement is neither a proposition nor a sentence (Foucault 1989a, 80, 82) and the rules governing statements are neither logical nor grammatical. Nor is it a speech act, which in fact depends on a network of statements (82–84). A statement is a kind of nonsense that allows sentences, propositions, and speech acts to "make sense" (86)—sense here understood as more than a matter of whether a proposition, sentence, or speech act is "well formed," since even incoherent propositions, grammatically incorrect sentences, and illegitimate speech acts may still have sense.[3] This makes it not a unit but "a function that cuts across a domain of structures and possible unities, and which reveals them, with concrete contents, in time and space" (87). Statements operate according to a principle of rarity different from that of propositions or sentences, one whereby, no matter how many statements exist, it is still not possible to say anything at any time and from any place: "statements (however numerous they may be) are always in deficit" (119). Statements always refer to other statements, their meaning determined not by any external set of rules but by this relational context.[4] This relation, however, is one that connects heterogeneous statements, bestowing sense on sentences, propositions, subjects, objects, and speech acts by establishing the necessary connections across divergent domains—for example, between medicine and law, which, as already seen, enables the propositions and sentences of psychopathology to make sense; or, as seen in Foucault's later works, between desire and truth, which underpins the various discourses of disciplinary society. A statement always refers beyond itself to an ensemble of other statements, but this ensemble is not simply exterior to the first statement but is rather folded into it internally and immanently.[5] This is why a statement, unlike a name, cannot be repeated—it "exists outside any possibility of reappearing" (89)—and yet "there can be no statement that in one way or another does not reactualize others" (98). It is also how a series of signs can become a statement only by virtue of "a

3. Compare with Lacan's (1981, 138–41) distinction between the level of enunciation (what is said) and the level of the statement (the sense of what is said).

4. "The meaning of a statement would be defined not by the treasure of intentions that it might contain, revealing and concealing it at the same time, but by the difference that articulates it upon the other real or possible statements, which are contemporary to it or to which it is opposed in the linear series of time" (Foucault 1989b, xix).

5. Deleuze (1988, 5) thus maintains: "A statement operates neither laterally nor vertically but transversally, and its rules are to be found on the same level as itself."

specific relation that concerns itself" (89), and yet "the referential of the statement forms the place, the condition, the field of emergence" (91) of subjects and objects, and the statement "cannot operate without the existence of an associated domain" (96) of other statements. The field of statements thus forms a complex and dispersed web, making the analysis of statements equivalent to that of discursive formations. The study of a discursive formation, Foucault says, is the study of its rarity, making it a study of both power and politics.

> In this sense, discourse ceases to be what it is for the exegetic attitude: an inexhaustible treasure from which one can always draw new, and always unpredictable riches; a providence that has always spoken in advance, and which enables one to hear, when one knows how to listen, retrospective oracles: it appears as an asset— finite, limited, desirable, useful—that has its own rules of appearance, but also its own conditions of appropriation and operation; an asset that consequently, from the moment of its existence (and not only in its "practical applications"), poses the question of power; an asset that is, by nature, the object of a struggle, a political struggle. (120)

The Archaeology does not develop a conceptualization of power, but it does contain themes that reappear in Foucault's later, genealogical work.[6] Beneath porous unities and identities lies a microscopic network of convergences, divergences, and disjunctions. The depth and solidity of these identities are simulations arising from the dynamics of discursive series. They are no less real for being semblances, but they do not have the substantiality often attributed to them, nor do they provide the stability for the discursive formations in which they appear. Only with the establishment of these stabilities can we speak properly of historical and chronological passage, but the dynamics of statements and discursive formations are not themselves

6. This point is often lost in interpretations (such as Dreyfus and Rabinow [1982, 102–5], and Habermas [1987, 266–70]) that hold The Archaeology to be a semistructuralist and linguistic-centered middle phase that Foucault rejected as he moved to an analysis of institutions and practices—which archaeological analysis bracketed off as "nondiscursive"—and therefore to an analytics of power acting on and through bodies. Foucault himself, however, never suggests that these "nondiscursive" elements are ever absent from or extraneous to the assemblages that constitute meanings and knowledges, and, as Deleuze notes (1988, 9–10, 31–33, 47–69), he soon abandons the term and its connotations of a strict separation from discursive formations.

chronological: "in its transformations, in its successive series, in its deriva-tions, the field of statements does not obey the temporality of the conscious-ness as its necessary model. . . . The analysis of statements operates therefore without reference to a cogito" (Foucault 1989a, 122). Discursive formations arise only with and through practices, which are socially and historically specific. But practices are always accompanied by a nonhistorical and untimely excess, which is neither a communally imposed opinion nor an anonymous and timeless truth that discourse partially uncovers, but rather an event-ness of the statement that is irreducible to the historicity of its associated practices.[7] A, Z, E, R, T—the alphabetical order adopted by French typewriters—has a specific spatiotemporal location: "it is always en-dowed with a certain materiality, and can always be situated in accordance with spatio-temporal coordinates" (86). A time and place marks its first use, and prior conditions explain its emergence. But who could reduce its sense and efficacy to this point of appearance, and how could it be treated as a mere historical contingency that might be changed at whim? The statement and its sense remain immanent to discursive formations and the practices but never localizable in them. It is this excess that gives discourse its pecu-liar discontinuity, making it an event that historical analysis has always tried to order in chronological sequence, "to preserve, against all decentrings, the sovereignty of the subject" (12). Power relations too are characterized by discontinuity and untimeliness even while they remain immanent to the practices that generate superficial identities and unities. They do not cause substantive unities to come into existence but rather condition the appear-ance of identities whose borders remain blurred and uncertain. Power rela-tions form a microscopic field of events that makes the analysis of power neither historical nor analytical but rather genealogical.

7. "In fact, it [the analysis of statements] is situated at the level of the 'it is said'—and we must not understand by this a sort of communal opinion, a collective representation that is imposed on every individual; we must not understand by it a great, anonymous voice that must, of necessity, speak through the discourses of everyone; but we must understand by it the totality of things said, the relations, the regularities, and the transformations that may be observed in them, the domain of which certain figures, certain intersections indicate the unique place of a speaking subject and may be given the name of author. 'Anyone who speaks,' but what he says is not said from anywhere. It is necessarily caught up in the play of an exteriority" (Foucault 1989a, 122).

12

The Genesis of the Surface I: The Theory of Drives

FREUD OFTEN LAMENTS that the understanding of instincts remains obscure, insufficiently established, and lagging in development compared to the rest of psychoanalytic theory (Freud 1953, 168n2; 1957a, 50; 1957b, 117–18; 1957c, 78; 1959, 56–57; 1965, 95; 1994, 45). Yet perhaps this is due primarily to the organization of component instincts into an oppositional schema that is extraneous and without basis. Freud defines instinct[1] as "an endosomatic, continuously flowing source of stimulation . . . lying on the frontier between the mental and the physical" (1953, 168, see also 1957b, 121–22). He contrasts it to "a 'stimulus,' which is set up by *single* excitations coming from *without*" (1953, 168) and to which the instincts respond (1959, 57). Instincts are only loosely related to particular objects (1953, 148) and their relations are further complicated by the ambivalence of desire, which amounts to a complex set of negative relations towards an object—the desire to possess, to destroy, to negate oneself and be possessed as an object, etc. (see Freud 1960, 29–30, 62, 157; 1961a, 41–42; 1994, 58–59). The largely unconscious resolution of these ambivalences through instinctual repression establishes both the precarious stability of the conscious self—"the inclusive unity of the ego" (1957a, 11)—and the processes of condensation, displacement, transference, and projection, which characterize unconscious life.

Instinct in general amounts to "a kind of elasticity of living things" (Freud 1959, 57) and the play of instincts, bringing together divergent do-

---

1. The vagaries of translation have meant that Freud's use of the German *trieb* has been consistently translated as "instinct," whereas Nietzsche translators have used both "instinct" and "drive." For the sake of consistency with the translations being quoted, I will continue to use the term "instinct" with respect to Freud, noting here Lacan's persuasive argument that the translation is poor.

mains and giving them meaning and direction, forms a surface of sense.[2] There is no necessary conceptual limit to the number of instincts—"in no region of psychology were we groping more in the dark. Everyone assumed the existence of as many instincts or 'basic instincts' as he chose" (Freud 1957a, 51)—yet Freud consistently restricts the component instincts to two—"our views have from the very first been *dualistic*" (53; see also 1994, 46). He does so, moreover, on an entirely speculative basis, from the speculations of ego-instincts operating against libido instincts (1957c, 75–81)[3] to the biological speculations of a death instinct beyond the pleasure principle (1957a).[4] Ultimately, Freud admits that the entire theory "is so to say our mythology. Instincts are mythical entities, magnificent in their indefiniteness" (1965, 95). Yet these binary schemes underpin the early and later models of the psyche, account for the ego's mechanisms of repression, which establish psychic life and its relation to the outside world, and remain useful hypotheses for explaining transference neuroses (1957b, 124) and, later, the compulsion to repeat witnessed in traumatic neuroses. It would seem, therefore, that a simplified and unconvincing understanding of the instincts is necessary for the kind of psychoanalytic theory Freud desires.

The great threat to multiplicity, for Freud, is not dualism but a collapse into monism, Jungian or otherwise (Freud 1957a, 52–53; see also 1966, 413).[5] The advantage of dualism is, purportedly, that the irreconcilability

2. Compare with Freud's definition of the sense of a psychical process: "We mean nothing other by it than the intention it serves and its position in a psychical continuity. In most of our researches we can replace 'sense' by 'intention' or 'purpose'" (1966, 40). Also: "By 'sense' we understand 'meaning,' 'intention,' 'purpose' and 'position in a continuous psychical context'" (61).

3. Indeed, the very existence of the ego and its instincts is displayed only indirectly through the failure and repression of libido instincts: "we are far less well acquainted with the development of the ego than of the libido, since it is only the study of the narcissistic neuroses that promises to give us an insight into the structure of the ego" (Freud 1966, 351). Also: "the analysis of the transference neuroses forced upon our notice the opposition between the 'sexual instincts' . . . and certain other instincts, with which we were very insufficiently acquainted and which we described provisionally as the 'ego instincts'" (1957a, 50–51). Freud is even more speculative when conceiving the opposition of Eros and Thanatos and theorizing the superego as a product of death-instincts, since "psycho-analysis has not enabled us hitherto to point to any [ego] instincts other than the libidinal ones" (53).

4. While the revised model of the psyche is not completed until *The Ego and the Id*, where, according to Freud "there are no fresh borrowings from biology, and on that account it stands closer to psycho-analysis than does *Beyond the Pleasure Principle*" (1961a, 12), Freud still admits that "if it were not for the considerations put forward in *Beyond the Pleasure Principle* . . . we should have difficulty in holding to our fundamental dualistic point of view" (46).

5. Freud's dualism, however, is limited compared with the multiplicity of earlier dynamic psychiatries. Ellenberger (1970, 145–47) holds Freud's development from the first to the second model of the psyche to be a move from dipsychism to polypsychism, yet the

of opposed instincts sustains the tension that propels psychic life. Yet the oppositional structure of the instincts falters in this regard and Freud finds himself led, in almost dialectical fashion, to the possibility of reconciliation. Thus, as Freud often notes in his later writings, the opposition between libido-instincts operating via the pleasure principle and ego-instincts operating by the reality principle proves unsustainable. Yet the later dualism between Eros and Thanatos is unable to furnish a "sharper expression" of "contrariety in instinctual life" (1965, 103).

The phenomenon of narcissism undercuts Freud's first instinctual opposition by locating libido-instincts in the ego, making the self-preservation instincts a modification of a more original self-love. As a result, "the distinction between the two kinds of instinct, which was originally regarded as in some sort of way *qualitative,* must now be characterized differently—namely as being *topographical*" (1957a, 52). Moreover, neither the ego nor the id really operates outside the pleasure principle: "the *reality principle* . . . does not abandon the intention of ultimately obtaining pleasure" (10). The condition of traumatic neurosis and the compulsion to repeat experiences that cannot be regarded as pleasurable, however, suggest another process at work. Through speculative analysis of the system *Pcpt.-Cs.* (perceptual-consciousness) and the child's *fort-da* game (14–17, 24–33), and drawing on Breuer's distinction between quiescent and mobile cathectic energies (26–27, 31), Freud theorizes the existence of free-flowing instinctual energies summoned to repair breaches in the psychic surface brought about by traumatic experiences. The repetition of disturbing experiences in traumatic neurosis is thereby explained as a mechanism for mastering overwhelming stimuli (32). These instinctual energies must be gathered together and cathected, first to the surface that separates the organism from the external world, forming a crust or resistant membrane that both receives and blunts external stimuli (26–27), and then to external objects of desire, before the pleasure principle can operate. Their flows obey primary psychical processes (34), whereas their binding represents a secondary process, which has precedence, "not . . . in *opposition* to the pleasure principle, but independently of it and to some extent in disregard of it" (35).

These "*freely mobile* processes which press towards discharge" are not in the first instance negatively related to any objects, as they "do not belong to the type of *bound* nervous processes" (1957a, 34). This idea comes close to a

---

tripartite organization of id, ego, and superego is certainly restricted compared with the plurality of subegos and personality clusters Ellenberger describes in earlier theories.

Nietzschean conception of drive as will to power[6] or Deleuze's definition of desire as "an *agencement* of heterogeneous elements that function" (Deleuze 1997a, 189). Freud, however, takes another route. Treating the compulsion to repeat as evidence of the fundamentally conservative nature of all instincts (1957a, 36), he claims that repetitions point to an instinct of living beings to return to an earlier state. However, to avoid any monism of desire, Freud establishes an opposition between different types of conservative instincts: sexual instincts that combine organic substances into larger unities (40–43) and other instincts that aim to return life to the inorganic.[7] This opposition cannot be fully subsumed under the binary of id and ego-instincts (53), leading Freud to locate a more primordial conflict between Eros and Thanatos. The existence of the death instinct remains merely a conviction (59; see also 1994, 46–47), only indirectly supported by the aggressiveness often observed together with the desire for an object (Freud 1957a, 54–55). Yet this presumption makes possible the final tripartite model of the psyche whose distinct components, once carved out of the mass of unconscious energies, are differentiated and consolidated as substantial structures by various oppositions and inversions: Eros turns the Thanatos outward, stopping what would otherwise be a self-destructive impulse but creating the ambivalence felt toward objects of desire; id and ego become modifications of Eros, the ego desexualizing and reversing sexual instincts in order to become "a coherent organization of mental processes" that "controls . . . the discharge of excitations into the external world" (1961a, 17) and the site of resistance to therapeutic efforts to recover the repressed (1957a, 19; 1961a, 17–18); and the death-instinct, turned outward but then forced back, forms the superego, which enforces social taboos with a repressive power exceeding the reality principle and, in melancholia, internalizes lost objects to form the ego-ideal (see Freud 1961a, parts 2–3). Deleuze's argument that Freud employs a restricted, material model of death as a merely inanimate state (Deleuze 1994, 111–12) was noted earlier. This new opposition between instincts, however, fails on its own terms and proves no better than that of the earlier model, when Freud admits that "if the pleasure principle had not

6. "Physiologists should think before putting down the instinct of self-preservation as the cardinal instinct of an organic being. A living thing seeks above all to *discharge* its strength—life itself is *will to power*; self-preservation is only one of the indirect and most frequent *results*" (Nietzsche 1989, §13).

7. Freud nonetheless laments that "we still feel our line of thought appreciably hampered by the fact that we cannot ascribe to the sexual instinct the characteristic of a compulsion to repeat" (1957a, 56). Compare also with Freud (1965, 107–8), where the conservative nature of erotic instincts' strivings to create larger unities remains an open question.

already been operative in [the primary psychic processes,] it could never have been established for the later ones," and that "the pleasure principle seems actually to serve the death instincts" (1957a, 63). Pleasure corresponding to a decrease in excitation (8), the conflict between Eros and Thanatos is really a struggle over ways to achieve this aim—which may in either case involve necessary unpleasurable aspects—yet this common aim was what made the opposition between id and ego-instincts untenable.[8]

Despite Freud's appraisal that Nietzsche's "guesses and intuitions often agree in the most astonishing way with the laborious findings of psychoanalysis" (Freud 1959, 60), Nietzsche's rejection of opposition precludes any substantive agreement between the two thinkers. Like Freud, Nietzsche posits a primordial struggle amongst conflicting drives—the self is thought to be a "multitude and disgregation of impulses" (Nietzsche 1968, §46) or "an inextricable multiplicity of ascending and descending life-processes" (§339). This self either lacks systematic order (corresponding to "weak will") or is coordinated by a dominant impulse ("strong will") (§46); its strength is assessed by its capacity to endure and overcome antagonism (§382). But as the workings of the drives remain obscure,[9] opposition is too blunt an instrument for approaching them. Nietzsche instead insists that the agonism of the drives be understood in terms of a more complex order of rank (§37). A world without essences or stabilities beneath appearances is "essentially a world of relationships" in which the sum of forces and resistances emanating from every point "is in every case quite incongruent" (§568). Understood in these terms, drives are relations that do not refer to prior, related terms, and their order of rank is not a fixed hierarchy but "a pattern of domination that *signifies* a unity but *is* not a unity" (§561).

Nietzsche's critique of mechanism is crucial to understanding this dynamic of drives. The mechanistic interpretation of the world, which "seems today to stand victorious" (1968, §618), abandons any assertion of purpose or goal in favor of a purely quantitative analysis. But this approach yields only abstract description and calculation, never explanation and comprehen-

---

8. Boothby (1991, 72–96, 109–10) interprets the pleasure associated with the death instinct as Lacanian *jouissance,* an inexpressible pleasure beyond the pleasure principle. However, Boothby also acknowledges (95–96) that this reading makes Freud's idea of two organically distinct classes of instincts untenable.

9. "However far a man may go in self-knowledge, nothing however can be more incomplete than his image of the totality of *drives* which constitute his being. He can scarcely name even the cruder ones: their number and strength, their ebb and flood, their play and counterplay among one another, and above all the laws of their *nutriment* remain wholly unknown to him" (Nietzsche 1982, §119).

sion (§§624, 660), and relies on undemonstrable forces of attraction and repulsion (§§620–21) and the hypothesis of the atom as the seat of motion (§§624–25). The mechanistic concept of force, therefore, "still needs to be completed: an inner will must be ascribed to it, which I designate as 'will to power,' i.e., as an insatiable desire to manifest power; or as the employment and exercise of power, as a creative drive, etc." (§619). The will to power satisfies the need to ascribe intentionality to force—"we cannot think of an attraction divorced from an *intention*" (§627)—but it cannot be conceived as a will in the ordinary sense (§692), nor as a cause that is separate from its effects: "That state of tension by virtue of which a force seeks to discharge itself—is not an example of 'willing'" (§668). To the degree that it can be aligned with causality, the will to power must be treated as a purely immanent cause: hence Nietzsche's declaration that "there is absolutely no other kind of causality than that of will [to power] upon will [to power]. Not explained mechanistically" (§658).

The link between quantity and quality allows the supplement of the will to power to be introduced immanently. On the one hand, "everything for which the word 'knowledge' makes any sense refers to the domain of reckoning, weighing, measuring, to the domain of quantity; while . . . all our sensations of value (i.e., simply our sensations) adhere precisely to qualities" (§565). On the other hand, "we need 'unities' in order to be able to reckon: that does not mean we must suppose that such unities exist" (§635). Mechanistic theory unreflectively adopts the abstraction of unity (along with the "sense prejudice" of motion in empty space), which explains the absolute divide it maintains between knowledge and value.[10] What the mechanistic account misses is *difference in quantity*, which follows from the idea of forces being essentially related and irreducibly unequal. The struggle of forces, being one of superior and inferior powers, introduces a qualitative element: "we cannot help feeling these differences in quantity as qualities" (§563).[11] These qualities, in turn, are "our perspective 'truths' which belong to us alone and can by no means be 'known'" (§565). Every force "adopts a

10. Compare with Nietzsche's call for another scientific project: "Our knowledge has become scientific to the extent that it is able to employ number and measure. The attempt should be made to see whether a scientific order of values could be constructed simply on a numerical and mensural scale of force" (1968, §710).

11. Also: "A greater power implies a different consciousness, feeling, desiring, a different perspective; growth itself is a desire to be more; the desire for an increase in quantum grows from a *quale*; in a purely quantitative world everything would be dead, stiff, motionless.—The reduction of all qualities to quantities is nonsense: what appears is that the one accompanies the other, an analogy" (Nietzsche 1968, §564).

perspective toward the entire remainder, i.e., its own particular valuation, mode of action, and mode of resistance" (§567). The will to power emerges, once the abstractions of mechanism are eliminated, as "a *pathos*" arising from "dynamic quanta, in a relation of tension to all other dynamic quanta" (§635). With it, the qualitative valuations of a dominating force's perspective are brought to the level of interpretation, the desire for discharge transforming the feeling of difference into a self-expression.

> The will to power *interprets* (—it is a question of interpretation when an organ is constructed): it defines limits, determines degrees, variations of power. Mere variations of power could not feel themselves to be such: there must be present something that wants to grow and interprets the value of whatever else wants to grow. Equal *in that*—In fact, interpretation is itself a means of becoming master of something. (The organic process constantly presupposes interpretations.) (§643)

The conflict among drives seeking only discharge, as a struggle of order of rank, is therefore one of quantitative differences in power that are simultaneously qualitative valuations: "Every drive is a kind of lust to rule; each one has its perspective that it would like to compel all the other drives to accept as a norm" (§481). A drive obtains its meaning and value only in relation to others: "In itself it has . . . neither this moral character nor any moral character at all, nor even a definite attendant sensation of pleasure or displeasure: it acquires all this, as its second nature, only when it enters into relations with drives already baptised good or evil or is noted as a quality of beings the people has already evaluated and determined in a moral sense" (Nietzsche 1982, §38; see also §35). At the same time, however, these evaluations are nothing more than the results of certain drives establishing dominance. The values they express are the "conditions of preservation and enhancement for complex forms of relative life-duration within the flux of becoming" and represent "the standpoint for the increase or decrease of these dominating centers ('multiplicities' in any case; but 'units' are nowhere present in the nature of becoming)" (Nietzsche 1968, §715). While the inequality of drives dictates that some drive dominates against myriad resistances, it is not a drive but a complex that rules. Or, rather, a drive, being a relation and not a substance, dominates by carrying out a seemingly enduring synthesis (endurance being only an appearance [§552]) of diverse and continually flowing forces. The relative stability of this synthesis results not from any law or equilibrium but rather from "the fact that . . . a certain force cannot be anything other than this certain force; that it can react to a quantum of resisting force only according to the measure of its strength" (§639). Thus, "a new arrangement of forces is achieved according to the measure of power of each of them" (§633). The dominance of this complex is extended by the evaluations it invests or cathects onto things so that, for example, it is not the feelings of pleasure and pain that underpin an evaluation of a thing but rather an evaluation that makes a thing pleasurable or painful (§260).

A series of falsifications, however, occurs in the becoming-dominant of a complex of drives. First, the multiple senses of any drive are reduced to an unambiguous meaning: "Every sovereign instinct . . . never lets itself be called by its *ugly* name. . . . All praise and blame in general crystallizes around every sovereign instinct to form a rigorous order and etiquette" (1968, §377). Moreover, drives tend to conceal themselves beneath their expression, as "the fleshy desires or the desires for power [hide] under the dominion of Christian values" (§377). Words, whose meanings are the products of victorious drives, each circulate a "weak emotion," which acts as "the common element, the basis of the concept" (§506), through the images and cases it brings together. Operating by similarity and identity, language can therefore grasp only the terminal and excessive outcomes of the drives, not the subtlety of their inner processes (1982, §115). The will to power's evaluations and judgments rest on processes of assimilation and equalization (§532). In all these ways, dominant drives tend to establish a simplified schema of fixed hierarchies, stable identities, and strict opposites—the categories of conscious life that are often erroneously applied to the conditions that generate consciousness (1968, §707). It is on this surface of identity and opposition that the illusion of an ego is projected.

Nietzsche is clear on this point: the ego, subject, or "I" is "only a fiction" (§370), "a perspective illusion" (§518) resulting from "our bad habit of taking a mnemonic, an abbreviative formula, to be an entity, finally as a cause" (§548). Similarly, "the will of psychology hitherto is an unjustified generalization . . . this will *does not exist at all*" (§692). These terms designate an epiphenomenon, which accompanies the dynamic of drives but is mistakenly taken as its unifying center: "The 'subject' is not something given, it is something added and invented and projected behind what there is" (§481). It is actually no more than a marker that aids in the "coordination and becoming conscious of 'impressions'" (§504). The thesis of the ego is thus a poor interpretation of the agency within the self, one that fails to recognize how "a deed often brings with it a numbness and lack of freedom: so that the doer is as if spellbound at its recollection and feels as if he were an *accessory* of it" (§235). What is seen to be the work of the ego combating a drive is really that of another drive, with the fictitious ego being nothing more than a spectator: "*that* one *desires* to combat the vehemence of a drive at all . . . does not stand within our own power . . . in this entire procedure our intellect is only the blind instrument of *another drive* which is a *rival* of the drive whose vehemence is tormenting us" (Nietzsche 1982, §109). Disregarding the duality between commanding and obeying impulses and thereby forgetting the order of rank, the regularity of sequences of thoughts and actions is treated not in terms of power relationships among forces but in terms of a coherent "I" that acts.

> Since in the great majority of cases there has been exercise of will only when the effect of the command—that is, obedience; that is, the action—was to be *expected*, the *appearance* has translated itself into the feeling, as if there were *a necessity of effect*. In short, he who wills believes with a fair amount of certainty that will and action are somehow one; he ascribes the success, the carrying out of the willing, to the will itself, and thereby enjoys an increase of the sensation of power which accompanies all success. (Nietzsche 1989, §19)

Like any dominant idea, that of the ego serves the purpose of dominant forces—"it could be useful and important for one's activity to interpret one-self *falsely*" (Nietzsche 1968, §492). Nevertheless, the ego "is *not* one with the central government of our nature" (§371). There is, indeed, no center of organization, formed and consolidated by opposition to what it is not. There is only the semblance of a center generated by a decentered system.

The principle grounding this entire dynamic and its accompanying illusions is the eternal return, the dissymmetrical structure of time and that which changes in time. As the being of becoming, the eternal return means that "every power draws its ultimate consequence at every moment" (Nietzsche 1968, §634) and thus that a complex process of order of rank and synthesis in the moment forms a cracked surface of sense. But more important, the eternal return demands that its repetition conceal itself beneath a simulation of identity, that dissymmetry mask itself as symmetry and continuity. By showing that identity and opposition have the status traditionally assigned to simulacra, the thought of eternal return opens the way to think beyond these terms. It thereby achieves the aim that Nietzsche sets for it: the reversal and revaluation of values.

13

The Genesis of the Surface II: Negation and Disjunction

FOR FREUD, negation is both a surface effect and what generates this surface. It is a correlate of the consciousness and intellect that make up the psyche's surface, but it is mapped more or less accurately onto the body's physical surface and the negative difference between inside and outside. Negation is thus a feature of reality and of the psyche's reality principle. No negation resides in the unconscious, where contraries condense into one another as though they were conformities (Freud 1966, 178), nor any linear time, with its spatialized difference between past, present, and future, nor fear of death, which is an abstract concept belonging to the mind's higher strata (Freud 1961a, 57–58). These are all products of the extension and withdrawal of cathectic energies to and from external objects (Freud 1961c, 231; 1995, 669) and of external frustrations and threats that, when they do not create certain pathologies where it collapses, establish the boundary line between the conscious portion of the ego and the outer world (Freud 1961a, 58–59; 1994, 2–4). Psychic conflicts are played out only on this boundary line, which provides them with sense and anchors them in real rather than phantasmatic origins. While the surface establishes a resonance between psychical and physical domains, the latter is privileged for the sake of Freud's aspiration to scientific rigor.

> The meaning of psychical conflict can be adequately expressed . . . by saying that for an *external* frustration to become pathogenic an *internal* frustration must be added to it. In that case, of course, the external and internal frustration relate to different paths and objects. The external frustration removes one possibility of satisfaction and the internal frustration seeks to exclude *another* possibility, about which the conflict then breaks out. I prefer this way of repre-

senting the matter because it has a secret content. For it hints at the
probability that the internal impediments arose from real external
obstacles during the prehistoric periods of human development.
(1966, 350)

Negation effects a compromise with repressed instincts and ideas, which
can enter consciousness only in the form of denials, although neither these
denials nor the patient's eventual acceptance of the repressed material re-
moves the processes of repression (Freud 1961b, 235–36). But negation is
also an expression of death instincts turned outward, as either a general
wish to negate found in certain psychoses or a symbol of negation that
grants thinking its independence from both the pleasure principle and repression (239). In thinking, negation assumes two roles corresponding to
the two functions of judgment: assigning attributes to subjects and deciding
the real existence of things. In assigning attributes, affirmation and negation correspond to oral processes of introjection and projection: "Expressed
in the language of the oldest—the oral—instinctual impulses, the judgment
is: 'I should like to eat this,' or 'I should like to spit it out'; . . . the original
pleasure-ego wants to introject into itself everything that is good and to eject
from itself everything that is bad. What is bad, what is alien to the ego and
what is external are, to begin with, identical" (237). Judgment of existence,
linked to the later reality-ego (237–38), similarly involves the inside/outside
division: "What is unreal, merely a presentation and subjective, is only internal; what is real is also there *outside*" (237). The ego and its boundaries,
however, are established and consolidated by two forms of real negation:
the withdrawal of objects and external threats. Through the temporary loss
and return of sources of excitation, "an *object* first presents itself to the ego
as something existing *outside*," while "a further stimulus to the growth and
formation of the ego, so that it becomes something more than a bundle of
sensations . . . is afforded by the frequent, unavoidable and manifold pains
and unpleasant sensations which the pleasure-principle . . . bids it abolish
or avoid" (Freud 1994, 3); through the castration complex, which functions
as a threat that dissolves the boy's Oedipus complex and a lack that initiates
the girl's penis envy, the identities of the sexes are defined and the higher
portions of the psyche—the ego-ideal and the superego—are secured (Freud
1961d; 1965, 112–35). Freud's different theorizations of neurotic anxiety are
also linked to these two forms of real negation, the early theory attributing
it to the loss of the mother rather than an extension of fear toward strangers
(1966, 406–7), and the later theory attributing it to a realistic (if not neces-

sarily real) castration threat (1965, 85–86, 93–94). All these real and psychic negations function throughout the stages of sexual development, in which libido instincts, once separated from their original link to vital functions, pass through oral, anal, and genital phases, organizing the body and its erotogenic zones and thereby coming to measure external reality in terms of the latter's promotion or hindrance of libido satisfaction (1966, chapters 20–21).

The death instinct serves as the internal source of negation, although it resides in the unconscious without manifesting any negative separation between contraries or otherwise expressing itself as such. Indeed, this internal impulse is actualized only through the denials and frustrations of the external world, allowing Freud to treat its negativity as derivative.[1] The death instinct works silently beneath the conspicuous and audible expressions of Eros (1961a, 40; 1994, 46–47) and, when turned outward and then back upon itself, establishes conscience and the superego. These final psychic negations exceed that of the reality principle, internalize the real negativity of external authority figures, and desexualize libido energies so that they can be sublimated into thinking (on this last point, see Freud 1961a, 45–46). But while they complete the structure of the psychic surface, Freud insists that these inversions, being themselves established at the surface, are linked to real and external prehistoric events. Conscience's original aggressions absorb renounced aggressions and use them to intensify feelings of guilt against the ego, but they also "represent a continuance of the rigour of external authority, and so have nothing to do with renunciation" (1994, 56). The issue thus arises concerning the origin of "the first quantum of aggressiveness with which the super-ego was endowed" (56). Freud considers two possibilities, the child, whose aggression is provoked by the external authority's denial of a desired object, and the authority figure itself, ultimately settling on the former, which accounts for the fact that children often

---

1. Freud, of course, holds that the death instinct is glimpsed only indirectly, making its existence a matter of speculation. In relation to negation and judgment, he therefore holds: "Affirmation—as a substitute for uniting—belongs to Eros; negation—the successor to expulsion—belongs to the instinct of destruction" (1961b, 239). Hyppolite (2006) holds that as a successor to expulsion, negation must be a distinct and subsequent development of the death drive, whose original processes can be pleasurable. Lacan (2006, 308–33) holds that Freud's expulsion refers to a primordial excision of oral impulses that goes beyond repression and constitutes the real, making negation a subsequent symbolization of the trace of this unsayable real. The issue of the existence of negativity and anxiety in the infantile unconscious became a central controversy between Klein and her opponents. On this controversy, see Kristeva (2001, 169–77) and Rose (1993, chapter 5), who both also discuss Klein's theories on negation and symbolization in relation to Lacan.

have strong superegos without experiencing strict upbringing (56–57; also 1965, 62, 109). But he also maintains that the child's being the source of unrenounced aggression fits with the reality of the primal patricide, since it indicates that some amount of infantile aggression must have been acted out rather than repressed (1994, 57–59).

Being civilization's greatest tool, it is easy to explain how the superego and its internalizations and identifications are reproduced across generations once civilization has developed. However, typical difficulties concerning fixed origins emerge with respect to the establishment of the first superego and the patricide and incest taboos that founded civilization. Prior to this, the ego's reality principle can create feelings of guilt associated with the dread of losing a loved object or being punished by an external authority (1994, 52–53). This "bad conscience"—which Freud admits is a misnomer[2]—is merely a modification of erotic instincts cathected to the ego, not a formation of the death instinct. Moreover, what transforms the primal patricide into the foundation of the superego is the brothers' remorse for committing their crime, but normal remorse "relates only to the one act, and clearly it presupposes that *conscience*, the capacity for feelings of guilt, was already in existence before the deed" (58). The transformation thus presupposes the superego it is meant to institute. Freud suggests another origin of remorse in the ambivalence the brothers necessarily felt toward their father. Once freed by the patricidal act, this remorse would found the superego.[3] A consciousness of guilt associated with aggressiveness would thus exist before conscience: "We ought not to speak of conscience before a super-ego is demonstrable; as to consciousness of guilt, we must admit that it comes into being before the super-ego, therefore before conscience" (63). Yet all this begs the question, for it remains unclear how consciousness of guilt, linked to remorse over a particular act, could generalize itself into conscience, which goes beyond the reality principle and polices acts and thoughts without distinction. The problem of circular origins now appears in the distinction between the consciousness of guilt and the sense of guilt,

2. "We call this state of mind a *bad conscience* but actually it does not deserve this name, for at this stage the sense of guilt is obviously only the dread of losing love, *social* anxiety" (Freud 1994, 52).

3. "This remorse was the result of the very earliest primal ambivalence of feelings towards the father: the sons hated him, but they loved him too; after their hate against him had been satisfied by their aggressive acts, their love came to expression in their remorse about the deed, set up the super-ego by identification with the father, gave it the father's power to punish as he would have done the aggression they had performed, and created the restrictions which should prevent a repetition of the deed" (Freud 1994, 58).

which Freud defines as "the severity of the super-ego" (63). Since the precise meaning of remorse is "a general term denoting the ego's reaction under a special form of the sense of guilt" (63),[4] the mechanism that leads remorse to conscience again presupposes the existence of conscience. Here Freud's attempt to ground the complex of negations composing the surface on a real and external event of negation reaches its limits.

Adopting his overall framework, Melanie Klein nonetheless inverts several of Freud's priorities in her account of the creation of the psychic surface. The superego's formation is linked to a series of pre-Oedipal developments, with an infantile superego corresponding to the infantile ego (Klein 1986, 63–64) and a sense of guilt carried by pregenital instinctual impulses (70). Object relations are constitutive, so that instead of an original narcissistic libido extending outward to become an object-libido (see Freud 1965, 102–3; 1966, 416), autoeroticism and narcissism result from the internalization of external good objects (Klein 1986, 204–5). It is therefore impossible to privilege external and real negations in psychic development: "external and internal situations are always interdependent, since introjection and projection operate side by side from the beginning of life" (53). Against Freud's thesis that anxiety has its source in external loss of the mother or castration threat, these are held to modify earlier anxiety situations involving the child's aggressiveness toward objects during the oral- and anal-sadistic phases (88, 92) and its early attempts to secure good objects.[5] The death instinct's negativity now has an original[6] and independent role in forming a surface of sense, which arises only when sadistic phantasies first relate the child to the external world (98) and then are modified so

4. The definitions of remorse, the consciousness of guilt, and the sense of guilt are all given after Freud suggests that "it may be just as well to go more precisely into the meaning of certain words like *super-ego, conscience, sense of guilt, need for punishment, remorse*, which we have perhaps often used too loosely and in place of one another" (1994, 63).

5. "The processes which subsequently become defined as 'loss of the loved object' are determined by the subject's sense of failure (during weaning and in the periods which precede and follow it) to secure his *good, internalized* object, i.e. to possess himself of it. One reason for his failure is that he has been unable to overcome his paranoid dread of internalized persecutors" (Klein 1986, 121).

6. "I do not share this [Freud's] view because my analytic observations show that there is in the unconscious a fear of annihilation of life. I would also think that if we assume the existence of a death instinct, we must also assume that in the deepest layers of the mind there is a response to this instinct in the form of fear of annihilation of life. Thus in my view the danger arising from the inner working of the death instinct is the first cause of anxiety. Since the struggle between the life and death instincts persists throughout life, this source of anxiety is never eliminated and enters as a perpetual factor into all anxiety-situations" (Klein 1975, 29).

that the accompanying anxiety does not overwhelm the ego or prevent the stabilization of symbolic relations (110). Conversely, schizophrenia results from "a severe inhibition of the capacity to form and use symbols, and so to develop phantasy life" (52). Phantasy too is therefore at the origin of the psychic surface: the infant first encounters a chaotic world of part-objects, from which "springs . . . the phantastic and unrealistic nature of the child's relation to all other things. . . . The object-world of the child in the first two or three months of its life could be described as consisting of hostile and persecuting, or else of gratifying parts and portions of the real world" (141). The initial resonance between the psychic and physical domains is thus one between phantasms and simulacra,[7] although normal developments, for Klein, create more stable boundaries between an integrated ego and an external world of whole objects.

In the infant's world, objects appear to be not only intermingled but merged together: "According to the child's earliest phantasies (or 'sexual theories') of parental coitus, the father's penis (or his whole body) becomes incorporated in the mother during the act" (Klein 1986, 96). Moreover, and reflecting the ambivalence of the child's sexual and aggressive instincts, these part-objects are seemingly intractable amalgamations of good and bad, making them both loved objects and hated objects. The objects being subject to introjection and projection, ambivalence resides both internally and externally. Real and phantastic objects double one another, creating a dialectic between inner and outer worlds that reinforces infantile anxieties.

> From the beginning the ego introjects objects "good" and "bad,"
> for both of which its mother's breast is the prototype—for good
> objects when the child obtains it and for bad when it fails him. But
> it is because the baby projects its own aggression on to these
> objects that it feels them to be "bad" and not only in that they
> frustrate its desires: the child conceives of them as actually danger-

---

7. This, at least, is Deleuze's term for the world of Klein's early infant: "We call this world of introjected and projected, alimentary and excremental partial internal objects the world of *simulacra*" (Deleuze 1990, 187). Kristeva, alternatively, speaks of it as a world of pseudo-objects or *abjects*: "because we cannot be certain of the identities that describe the archaic link between the ego and the Other, it may be more helpful to speak of an *abject* rather than of an ego or an object already there. The future subject is founded upon a dynamic of abjection whose optimal quality is fascination. And if this future subject readily grants himself a 'presence' of other people that he internalizes as much it [sic] expels, he is not facing an object but, in fact, an *ab-ject*, with this *a* understood in the privative sense of the prefix, that is, as vitiating the object as well as the emerging subject" (Kristeva 2001, 72–73).

ous—persecutors who it fears will devour it, scoop out the inside of its body, cut it to pieces, poison it—in short, compassing its destruction by all the means which sadism can devise. These imagos, which are a phantastically distorted picture of the real objects upon which they are based, are installed by it not only in the outside world but, by the process of incorporation, also within the ego. Hence, quite little children pass through anxiety situations (and react to them with defence mechanisms), the content of which is comparable to that of the psychoses of adults. (116–17)

Responding like a tiny Manichean, the infant obsessively tries to cleanly separate good and bad. This splitting practice reflects both the initial weakness of the ego, which functions to integrate differences gradually, and the powerful negativity of the death instinct, which the infantile ego must thrust outward to defend itself against primordial anxiety (216). The aim of separating good and bad initiates processes of introjection and projection. But the barriers established by splitting continually falter and their contents seep into one another. Bringing its sadistic impulses to bear on the good and bad objects it tries to separate, the child only reinforces its own persecutory fears that its actions will bring about reprisals. Owing to the connection between inside and outside, the ego's attempts to split off good and bad part-objects also divide and weaken it (183–86); yet when it consolidates itself by introjecting good objects, the ego necessarily absorbs bad objects as well (181). The infant endures a paranoid-schizoid position, characterized by idealizing the good object and seeking its protection (202). As attempted separations repeatedly fail, the infantile superego's demands on the ego intensify.

The ego endeavours to keep the good apart from the bad, and the real from the phantastic objects. The result is a conception of extremely bad and *extremely perfect* objects, that is to say, its loved objects are in many ways intensely moral and exacting. At the same time, since the ego cannot really keep its good and bad objects apart in its mind, some of the cruelty of the bad objects and of the id becomes related to the good objects and this then again increases the severity of their demands. These strict demands serve the purpose of supporting the ego in its fight against its uncontrollable hatred and its bad attacking objects, with whom the ego is partly identified. The stronger the anxiety is of losing the loved objects, the more the ego strives to save them, and the harder the task of restoration becomes the stricter will grow the demands which are associated with the super-ego. (123)

The ego's development makes it better able to integrate and synthesize good and bad part-objects (203). Moreover, "as the adaptation to the external world increases . . . splitting is carried out on planes which gradually become increasingly nearer and nearer to reality" (144). The ego's unification, the consolidation of realistic divisions between inside and outside and between objects, and the integration of part-objects into complete objects (on the last issue, see 118), are thereby interconnected. Early schizophrenia recedes with the beginnings of a stable surface, and sexual instincts, establishing independence from and ascendancy over death instincts, begin organizing the child's body and erotogenic zones. But these steps create a new, manic-depressive position, characterized by the loss of the loved object and delusions of omnipotence in which the child denies the existence of bad objects and envisions dominating objects in general (132–34, 182).

> The ego's growing capacity for integration and synthesis leads more and more . . . to states in which love and hatred, and correspondingly the good and bad aspects of objects, are being synthesized; and this gives rise to the second form of anxiety—depressive anxiety—for the infant's aggressive impulses and desires towards the bad breast (mother) are now felt to be a danger to the good breast (mother) as well . . . at this stage the infant increasingly perceives and introjects the mother as a person. Depressive anxiety is intensified, for the infant feels he has destroyed or is destroying a whole object by his greed and uncontrollable aggression. (203)

On the one hand, "the more integrated ego becomes capable of experiencing guilt and feelings of responsibility which it was unable to face in infancy" (228). On the other hand, "the very experience of depressive feelings . . . has the effect of further integrating the ego, because it makes for an increased understanding of psychic reality and better perception of the external world, as well as for a greater synthesis between inner and external situations" (189). A stronger ego, able to acknowledge and control its destructive side, is more capable of love (227), and the child, acting from both love and guilt, sets out to repair its damaged love object and regain the latter's love. The choice of love object is now determined in the passage to the genital phase of sexual development. The boy, whose libido changes to the aim of penetration, retains the mother as his primary object, while the girl, keeping the receptive aim of the earlier oral phase, turns to the father (70). The desire to repair the good object, fostered by the depressive position, thereby initiates the Oedipus complex, which, as with Freud, completes the psychic surface's development.

Normal development, for Klein as for Freud, culminates in the relative adequacy of psychic life to the real separations between inside and outside and between real external objects. Klein, for example, maintains that in passing through persecutory and depressive positions, "anxieties lose in strength; objects become both less idealized and less terrifying, and the ego becomes more unified. All this is interconnected with the growing perception of reality and adaptation to it" (1986, 189–90). When Deleuze uses Klein to theorize the construction of the surface, he takes issue with this very point: the child never outgrows the chaos of part-objects, he holds, because reality is a simulacrum—a disjunction of differences folded into one another and linked through a differenciator, dark precursor, or pure event. As such, the phantasm, relating to external reality at and through the surface, cannot be a good or bad copy of this reality. The issue of whether phantasms have real or imaginary causes is thereby displaced.[8] Nevertheless, if the child strives to separate good and bad part-objects, where among the simulacra of part-objects does it find its inspiration? Here Deleuze maintains that since good is associated with purity and fullness, all part-objects are necessarily bad: "every piece is bad in principle (that is, persecuting and persecutor), only what is wholesome and complete is good" (Deleuze 1990, 188). The principle of purity must therefore come from the differenciator—here termed the "body without organs"—which, univocally circulating through the dispersed flows, gives the appearance of a pure object that the child can use as a center around which to organize itself and its surroundings (187–88). A simulation of a transcendent reference point emerging from the simulacrum thereby enables the move from the paranoid-schizoid to the depressive position.

---

8. "Freud was then right to maintain the rights of reality in the production of phantasms, even when he recognized them as products transcending reality. It would be unfortunate if we were to forget or feign to forget that children do observe their parents' bodies and parental coitus; that they really become the object of seduction on the part of adults; that they are subjected to precise and detailed threats of castration, etc. Moreover, parricide, incest, poisoning, and eventration are not exactly absent from public and private histories. The fact is, though, that phantasms, even when they are effects and because they are effects, differ in nature from their real causes" (Deleuze 1990, 210–11).

What the schizoid position opposes to bad partial objects—introjected and projected, toxic and excremental, oral and anal—is not a good object, even if it were partial. What is opposed is rather an organism without parts, a body without organs, with neither mouth nor anus, having given up all introjection or projection, and being complete, at this price. . . . *Insofar as it is the principle of the depressive position*, the good object is not the successor of the schizoid position, but rather forms itself in the current of this position, with borrowings, blockages, and pressures which attest to a constant communication between the two. (188, 190)

Characterized as transcendent and enigmatic, "the good object is by nature a lost object. It only shows itself and appears from the start as already lost, as *having been lost*" (191). Consigned to the heights, it progressively organizes the child's erotogenic zones, coordinating the child's part-objects to complete the bodily surface. This process culminating in the genital phase of sexual development, the phallus becomes a privileged signifier, serving "the direct and global function of integration, or of general coordination" (200). The phallus differs from both the good object of the heights and the simulacra of bad part-objects of the depths. Although donated by the good object—"the child receives the phallus as an image that the good ideal penis projects over the genital zone of his body" (203)—it belongs wholly to the surface, aiming to mediate its connection to the heights. It is "an instrument of the surface, meant to *mend* the wounds that the destructive drives, bad internal objects, and the penis of depths have inflicted on the maternal body, and to reassure the good object, to convince it not to turn its face away" (201). It gives "to the child's penis the force of embarking on the venture" (206) of making reparations to the damaged object.

The boy's loving attempt to use the phallus to repair his mother, which involves substituting himself for his father, thus begins the Oedipal complex. With the ego's consolidation in the depressive position, sexual instincts are separated from and raised above death instincts. But aggressive impulses remain operative underneath, and the phallus's role in defining sexual difference brings the castration complex center stage in specifying the physical surface of each character in the drama. Libido instincts are repressed, desexualized, and sublimated to form "the second screen, the cerebral or metaphysical surface" (218). Death, castration, and murder become never fully identifiable elements of an Oedipal phantasm circulating between the physical surface of sexuality and this metaphysical surface

of thought. Sexuality is brought into thought, as the trace of castration remains even after sexual energies have been sublimated; thought, via symbolization, reinvests its desexualized energies on the body's surface
(242–43). The surfaces fold into each other but remain irreducible: the sexual organization of the physical surface prefigures language (230–33, 241–
42), but language arises only insofar as sexuality is sublimated into
something different; symbolization, in turn, can never collapse the symbol
into what is symbolized. The Oedipal phantasm resonates between these
sexual and desexualized surfaces, but it also refers to pregenital and genital
sexual series or organizations, which it causes to resonate in the unconscious, so that "the phantasm requires four series and two movements"
(240).

The surface is thereby composed of disjunctions upon disjunctions: the
disjunctions of simulacra produce the appearance of the lost good object
that separates sexual and aggressive drives and organizes pregenital and
Oedipal sexual surfaces; the disjunction of these surfaces desexualizes and
sublimates libido energies into thought; and the surfaces of sexuality and
thought continue to resonate through their difference. The pregenital organization of erotogenic zones, the Oedipal organization around the phallus,
and the post-Oedipal resonance of sexual surfaces follow the order of connective, conjunctive, and disjunctive syntheses that Deleuze outlines in relation to the logic of sense (224–27). The disjunctive movement of the
phantasm is the nonsense that generates sense, underpinning the identities
of denoted bodies, signified concepts, and the self that manifests or expresses itself in language. But precisely because it is a movement of disjunction, the phantasm is also linked to a decentering through which these
identities, and particularly the identity of the ego, are dissolved. Although
the phantasm "finds its point of departure (or its author) in the phallic
ego of secondary narcissism," seeming to depend on the ego's pre-Oedipal
consolidation, within the phantasm the ego "is neither active nor passive
and does not allow itself at any moment to be fixed in a place, even if this
place were reversible" (212). This dissolution of the ego, Deleuze argues,
must not be confused with a similar dissolution carried out in dialectics
(213). Arising through disjunction, it transforms the ego into an event: "the
individuality of the ego merges with the event of the phantasm itself, even
if that which the event represents in the phantasm is understood as another
individual, or rather as a series of other individuals through which the dissolved ego passes" (213–14).

If a phantasm is an event that "transcends inside and outside, since its topological property is to bring 'its' internal and external sides into contact, in order for them to unfold onto a single side" (211), then the Oedipal phantasm is the ultimate Event to which various surface events, differing from themselves and from the diverse series they bring together, refer. Castration embeds itself as a crack in both physical and metaphysical surfaces, an enigmatic "something is there" effecting their communication on a univocal plane.[9] In this way, the Oedipal phantasm becomes "the site of the eternal return" (220), the out-of-sync structure of that which changes. The resonance of two series through a differenciator is the "intrinsic beginning" of the phantasm, but "the phantasm develops to the extent that the resonance induces a *forced movement* that goes beyond and sweeps away the basic series . . . the forced movement of an amplitude greater than the initial movement" (239). In this regard, while Eros's sexual instincts initiate the surface's development, it is Thanatos, the death instinct, that performs the excessive forced movement, dissolving the ego, desexualizing libido instincts, and sublimating them into thought: "We can therefore name the entire forced movement 'death instinct,' and name its full amplitude 'metaphysical surface'" (240). With this, death is released from Freud's material model and becomes the positive force of repetition/contraction: "Thanatos (as transcendental principle) is that which gives repetition to Eros, that which submits Eros to repetition" (Deleuze 1994, 18). Raising disjunction to the level of thinking, this positivity makes way for a creative break with the compulsions and necessities of both the instincts and the past—hence Deleuze links the eternal return to the future's openness.[10] There is, of course, uncreative thinking—"the risk is obviously that the phantasm falls back on the poorest thought, on a puerile and redundant diurnal reverie 'about' sexuality" (Deleuze 1990, 220)—but when the phantasm sustains the resonance between sexuality and thought, it finds its "path of glory" in

9. "The trace of castration as a deadly furrow becomes this crack of thought, which marks the powerlessness to think, but also the line and the point from which thought invests its new surface. And precisely because castration is somehow between two surfaces, it does not submit to this transmutation without carrying along its share of appurtenance, without folding in a certain manner and projecting the entire corporeal surface of sexuality over the metaphysical surface of thought. The phantasm's formula is this: from the sexual pair to thought via castration" (Deleuze 1990, 218).

10. "The ultimate synthesis concerns only the future, since it announces in the superego the destruction of the Id and the ego, of the past as well as the present, of the condition and the agent. . . . If there is an essential relation between eternal return and death, it is because it promises and implies 'once and for all' the death of that which is one. If there is an essential relation with the future, it is because the future is the deployment and explication of the multiple, of the different and of the fortuitous, for themselves and 'for all times'" (Deleuze 1994, 115).

the thought of eternal return: "What kind of metamorphosis is it, when thought invests (or reinvests) that which is projected over its surface with its own desexualized energy? The answer is that thought does it in the guise of the Event" (220). Raised to this level, thinking becomes inseparable from transmutation and the revaluation of values.

14

Crisis Time: Nihilism and the Will to Truth

FOR NIETZSCHE, modern nihilism is a condition in which delegitimated values of the past remain embedded in an incompatible present. Morality is "a system of evaluations that partially coincides with the conditions of a creature's life" (Nietzsche 1968, §256) because "feelings about values are always behind the times; they express conditions of preservation and growth that belong to times long gone by; they resist new conditions of existence with which they cannot cope and which they necessarily misunderstand" (§110). In its out-of-sync character, nihilism expresses more generally the formal structure of time. But it arises only when realization of this condition combines with a failure of will and an inability to find meaning and sense in this time out of joint.[1] Thus "it is in one particular interpretation, the Christian-moral one, that nihilism is rooted" (§1), and its advent occurs when the failure to sustain this interpretation prompts the conclusion that the world is meaningless: "the untenability of one interpretation of the world, upon which a tremendous amount of energy has been lavished, awakens the suspicion that *all* interpretations of the world are false" (§1). As an expression of weak or slavish will, nihilism is at once a historical culmination and the foundation of human history and culture as such: it is both historical and genealogical.[2] But the historical moment, announced in the

---

1. Nietzsche (1968, §95) thus considers the nineteenth century to be "more animalic and subterranean, uglier, more realistic and vulgar, and precisely for that reason 'better,' 'more honest,' more submissive before every kind of 'reality,' truer; but weak in will." See also 1974, §347, where nihilism is linked to faith, which "is always coveted most and needed most urgently where will is lacking."

2. On the one hand: "For why has the advent of nihilism become *necessary?* Because the values we have had hitherto thus draw their final consequence; because nihilism represents the ultimate logical conclusion of our great values and ideals" (Nietzsche 1968, preface, §4). On the other hand: "Supposing that what is at any rate believed to be the 'truth' really is true, and the *meaning of all culture* is the reduction of the beast of prey 'man' to a tame and civilized

madman's declaration of God's death, in no way marks a linear break or progression. "God is dead; but given the way of men, there may still be caves for thousands of years in which his shadow will be shown" (Nietzsche 1974, §108); the madman necessarily comes too soon, giving rise to the paradox of an event that has already taken place and yet is still to come.[3]

Considered historically, as a specifically modern condition, nihilism emerges from a skepticism and sentimentality that follows the European Enlightenment (Nietzsche 1968, §§91, 96). Its roots are in a subtle and innocuous shift tied to the age of reason: the separation of the truth of the world from the idea that God created it, a move that places truth on the side of science and God on the side of belief. After this move, it is still possible to believe in both God and truth, but even within this continuity they shine in a world that has lost significant color. The more splendid colors with which the world was painted when a god was seen to shine in it are dulled,[4] but our modern lack of "historical sense"—the failure to appreciate the heterogeneity of the past without reducing it to familiar representations[5]— prevents us from noticing the difference. On a superficial level, continued belief in God seems compatible with enlightenment: the empirical evidence hardly supports the madman's proclamations that we have "wipe[d] away the entire horizon," "unchained this earth from its sun," and sent ourselves

---

animal, a *domestic animal*, then one would undoubtedly have to regard all those instincts of reaction and *ressentiment* through whose aid the noble races and their ideals were finally confounded and overthrown as the actual *instruments of culture*" (Nietzsche 1967, 1.11). See also Deleuze (1983, 34) and Heidegger (1977a, 53–112), although Heidegger, as part of a larger attempt to show that Nietzsche remains ensconced in metaphysics, holds that Nietzsche, not the advocates of Christianity, is the nihilist.

3. "'I have come too early,' he [the madman] said then; 'my time is not yet. This tremendous event is still on its way, still wandering; it has not yet reached the ears of men. Lightning and thunder require time; the light of the stars requires time; deeds, though done, still require time to be seen and heard. This deed is still more distant from them than the most distant stars—*and yet they have done it themselves*'" (Nietzsche 1974, §125).

4. "The illumination and the color of all things have changed. We no longer understand altogether how the ancients experienced what was most familiar and frequent—for example, the day and waking. Since the ancients believed in dreams, waking appeared in a different light. The same goes for the whole of life, which was illumined by death and its significance; for us 'death' means something quite different. All experiences shone differently because a god shone through them. All decisions and perspectives on the remote future, too; for they had oracles and secret portents and believed in prophesy. . . . We have given things a new color; we go on painting them continually. But what do all our efforts to date avail when we hold them against the colored splendor of that old master—ancient humanity?" (Nietzsche 1974, §152).

5. See Nietzsche (1974, §§83, 337; 1983, 60–63). On the false historical sense of modern Germans, see Nietzsche (1974, §377). Nietzsche is particularly dismissive of the lack of historical sense of English utilitarian psychologists. See Nietzsche (1967, 1.2).

"plunging continually. . . . Backward, sideward, forward, in all directions. . . . Is there still any up or down?" (Nietzsche 1974, §125). But, like Wile E. Coyote chasing the Roadrunner, we have run off the cliff and simply not yet looked down. His anxiety as he places his paw beneath the cloud of smoke in order to feel whether there is ground beneath him is very much our own.

The madman enters the marketplace seeking God, encounters laughter from nonbelievers, and only then declares that God is dead (Nietzsche 1974, §125). The madman is no longer the Shakespearean fool speaking a mysterious truth that others ignore at their peril. Truth and God are thereby separated: " 'Truth' was experienced differently, for the insane could be accepted formerly as its mouthpiece—which makes *us* shudder or laugh" (§152). It is not that madness now lacks truth. Rather, its truth is not a matter of the madman, like all things, being God's creation. Whatever truth now attaches to madness comes from its being "medicalized" and renamed "insanity," an object of scientific study and treatment. Ironically, laughing at the madman is the sign that the enchanted world of signs, where madness spoke divine truth, is gone.[6]

God is dead, but the full import of the event remains unknown (§343). His legacy lives on in the rationality that seems to have dismissed him: hence Nietzsche's claim that science is not truly opposed to the religious ascetic ideal (see Nietzsche 1967, 3.23–25). Truth, now reconfigured as scholarship or science, still carries residues of the theological and metaphysical conceptions it claims to surpass, residues appearing in science's continuing reliance on merely aesthetic judgments: that truth is simple; that it is pure; that the truth of something is that which remains constant over time, unchanging and universal; that truth is good, and that what is good is true. These valuations, which equate goodness with purity, universality, truth, usefulness, and beauty, similarly linking evil to the opposite terms, have simply been transferred from the transcendent suprasensory realm of Platonism and Christianity to modern ideals of objectivity, natural law, and purity of method. But they remain mere convictions, and science can never raise itself above the level of mere belief because its values depend on judgments that it dismisses as mere beliefs.

Modern science thereby continues the will to truth that drove its predecessors. The will to truth does not seek after truth but instead demands that

---

6. On this last point, see Connolly (1993, 7–9). On the shifts in the treatment of madness from being a mysterious truth to a public scandal, an error of reason, a moral failing, and finally a medical condition open to study and discipline, see, of course, Foucault (1989c).

the world conform to ideals of purity and universality associated with a particular conception of truth, ideals that cannot sustain themselves once they are disconnected from the divine source that grounded them. Nietzsche offers two possible interpretations of science's will to truth: as "the will *not to allow oneself to be deceived*" or as "the will *not to deceive*" (1974, §344). The former is consistent with a cautious pragmatism—"one does not want to allow oneself to be deceived because one assumes that it is harmful, dangerous, calamitous to be deceived" (§344)—but its reasoning is not unconditional: one might not want to be deceived but also accept that at times it is better not to know the truth. But the unconditional demand for truth, which cannot be justified by actual worldly experience,[7] "must have originated *in spite of* the fact that the disutility and dangerousness of 'the will to truth,' of 'truth at any price' is proved to it constantly" (§344). This conviction in favor of truth therefore rests on nothing more than a moral demand that truth is good[8] and concomitant aesthetic judgments that what is good is also pure, simple, and universal. Thus "it is still a *metaphysical faith* upon which our faith in science rests," and the demand not to deceive has an ambiguous status in a world where life aims "at semblance, meaning error, deception, simulation, delusion, self-delusion, and when the great sweep of life has actually always shown itself to be on the side of the most unscrupulous *polytropoi*" (§344). We simply need to entertain seriously the possibility that truth is not always good, that evil is sometimes useful, and that purity and endurance are merely simulations, in order to see the danger of this dogmatic stance.

It does little to alleviate the situation that many scientists willingly admit the uncertainty of their knowledge or adopt some antifoundationalist posture. What characterizes nihilism is not that belief in these ideals is gone, but that it continues surreptitiously in an incompatible world, with no consideration of how the loss of foundations requires a revaluation of values.[9]

7. "Is wanting not to allow oneself to be deceived really less harmful, less dangerous, less calamitous? What do you know in advance of the character of existence to be able to decide whether the greater advantage is on the side of the unconditionally mistrustful or of the unconditionally trusting? But if both should be required, much trust *as well as* much mistrust, from where would science then be permitted to take its unconditional faith or conviction on which it rests, that truth is more important than any other thing, including every other conviction? Precisely this conviction could never have come into being if both truth and untruth constantly proved to be useful, which is the case" (Nietzsche 1974, §344).

8. "Consequently, 'will to truth' does *not* mean 'I will not allow myself to be deceived' but—there is no alternative—'I will not deceive, not even myself'; *and with that we stand on moral ground*" (Nietzsche 1974, §344).

9. "Much less may one suppose that many people know as yet *what* this event really means—and how much must collapse now that this faith has been undermined because it

Just as belief in God continues after his death, the old values are continually espoused, although they remain mere espousals that, perhaps, even their advocates no longer really believe. Thus, even while modern life finds numerous defenders, the melancholy of the times continues to be felt. This is apparent in the way technological progress is celebrated alongside worry about increasing environmental damage, the way democracy is declared the best political system despite routine lamentations that all politicians are the same, the way faith in human progress continues even after Auschwitz and the bomb, and the way the end of history has been announced, with liberalism having seen off its evil enemy, and yet no one seems happier. In all these cases purity, progress, and truth are idealized even while it becomes harder to take them seriously. Yet at the same time, these ideals are put forward even more severely. Hence Nietzsche's concern with the "bovine nationalism" (1968, §748) rising throughout Europe, reflecting a herd complex all the more stringent because it sees itself under threat, and his prediction that modernity's leveling forces will create a system of global economic management in which "mankind will be able to find its best meaning as a machine in the service of this economy—as a tremendous clockwork, composed of ever smaller, ever more subtly 'adapted' gears" (§866).

The will to truth's continuing demands act as a police mechanism on modern discourse. As a porous pseudounity arising in the intersections of divergent discourses, a discourse always exceeds the loose unity of its objects, the shifting authority of its subjects, and the ephemeral legitimacy of its knowledges. Foucault thus holds that "in every society the production of discourse is at once controlled, selected, organized and redistributed by a certain number of procedures whose role is to ward off its powers and dangers, to gain mastery over its chance events, to evade its ponderous, formidable materiality" (1984b, 109). Foucault further contends that while there are "external" restrictions on discourse, such as prohibition and rejection, today an "internal" limitation, the will to truth, "increasingly attempts to assimilate the others, both in order to modify them and to provide them with a foundation" (113). The restrictive aspects of the will to truth are ambiguous, since they also enable the production of discourses. The division of texts into primary and commentary, the use of the author principle to ground the unity of a group of texts, and the formation of disciplines that delineate domains of objects, methods of study, and principles for testing

---

was built upon this faith, propped up by it, grown into it; for example, the whole of our European morality" (Nietzsche 1974, §343).

the validity of propositions (see 114–20) all enable discourses and "true statements" to proliferate endlessly, but always within controlled conditions. They suggest both that creation is possible within discourses and that such new beginnings have always already been ordered so that nothing in discourse is really dangerous.[10] As such, they betray anxiety over excess.

The will to truth effects historically contingent divisions between truth and falsity and between power and true discourse: whereas in an ancient conception true discourse carried the power of inspiration and domination—it was the discourse of the strong—from Socrates and Plato onward true discourse becomes the discourse free of such power effects.[11] Whatever force true discourse now carries comes from nothing other than its truth, so that the will to truth becomes an internal limitation by compelling discourse to tell the truth about itself independent of any attachment to power or a powerful speaker. In this way, however, it hides the exercise of power that still flows through it, a power that follows from discourse being driven by a *will* to truth: "'True' discourse, freed from desire and power by the necessity of its form, cannot recognize the will to truth which pervades it" (114). Yet, ironically, the will to truth, being a demand that discourse conform to the ideal of purity, also licenses great self-deception and ignorance. As Foucault argues, the content of nineteenth-century discourses on human sexuality are feeble "from the standpoint of elementary rationality, not to mention scientificity" (1990a, 54). These discourses are formed in the intersection of a biological discourse of reproduction and a medical discourse of sex, both of which employed "quite different rules of formation" and related to each other with "no real exchange, no reciprocal structuration" (54–55). These disparities "indicate that the aim of such a discourse was not to state

10. Hence Foucault opens "The Order of Discourse" by suggesting both that he wishes for a stable tradition behind him and upon which he could draw, so as not to have to begin discourse himself, and that the institution assures that beginnings never pose a problem: "Desire says: 'I should not like to have to enter this risky order of discourse; I should not like to be involved in its peremptoriness and decisiveness; I should like it to be all around me like a calm, deep transparence, infinitely open, where others would fit in with my expectations, and from which truths would emerge one by one; I should only have to let myself be carried, within it and by it, like a happy wreck.' The institution replies: 'You should not be afraid of beginnings; we are all here in order to show you that discourse belongs to the order of laws, that we have long been looking after its appearances; that a place has been made ready for it, a place which honours it but disarms it; and that if discourse may sometimes have some power, nevertheless it is from us and us alone that it gets it'" (1984b, 109).

11. "Between Hesiod and Plato a certain division was established, separating true discourse from false discourse: a new division because henceforth the true discourse is no longer precious and desirable, since it is no longer the one linked to the exercise of power. The sophist is banished" (Foucault 1984b, 112).

the truth but to prevent its very emergence" (55). The will to truth therefore functions only to give certain forms of discourse and knowledge an authoritative stamp of truthfulness or scientificity, which is quite compatible with willing various sorts of nonknowledge: "Choosing not to recognize was yet another vagary of the will to truth" (55).

The modern will to truth, as Foucault's genealogical works detail, seeks to define and delineate various forms of deviancy and delinquency in order to better police standards of normality and compel individuals toward these norms against opposing resistances. In other words, it aims to secure the purity of an identity deemed good and healthy against identities defined as evil or sick, thereby expressing binary valuations comparable to those of earlier times. The explosion of discourses surrounding sexual acts falling outside the norm, the more precise separations between criminal acts where responsibility can be assigned and acts that signal psychological deficiencies, the increasingly detailed definitions and classifications of abnormalities, and the intensification of strategies of surveillance, testing, and confession all relate to this goal (see Foucault 1990a, 36–49). Nevertheless, this aim of the will to truth must not be confused with the dynamic relations of power, resistance, dispersion, and disjunction that are both deployed by this will and from which it emerges. The strategies of the will to truth must not be confused with the efficacy of power per se, because this reduces this efficacy to a dynamic comprehensible to this same, problematic will to truth.

If the will to truth can be traced to Plato, then its error can be understood as more than its claiming that true discourse is free of power. In disparaging the world of becoming, Plato locates truth in the realm of timeless Forms. He thereby invests truth with a positive and transcendent identity and understands difference in terms of falling away from this identity into its opposite, so that ugliness is no more than a lack of Beauty and the realm of becoming is a mixture of Being and its opposite, nothingness. In this respect, Plato's error, like the error of traditional metaphysics more generally, is to mistake a simulation—identity—for substantial truth, while measuring difference in terms of degrees of opposition to this identity. At the same time, what problematizes this substantiality by being neither truth nor its opposite—what Plato associates with art, acting, and sophistry—is denigrated as a mere simulation of the truth and the true discourse of the philosopher. Thus the will to truth not only sees truth as being free of power but also conceives of its world in terms of identity and opposition. It irons out the discontinuity and divergence that archaeological analysis uncovers, unfolding these into a difference understood in terms of inside and outside.

Under the conditions of a general demand for secure identity, modernity, Foucault holds, sees the rise of a "tricky combination . . . of individualiza-tion techniques, and of totalization procedures" (1982, 213). On the one hand, an originally Christian pastoral form of power now seeks knowledge of the most intricate details of individuals under the auspices of ensuring their prosperity in this world. This task is inseparable from "knowing the inside of people's minds" (214), making confession and case study impor-tant mechanisms of this power. On the other hand, increasingly strict stan-dards of normality are enforced by ever proliferating institutional agents, who police the division between subject and nonsubject. People are com-pelled to conform, and individualization is a means to this goal. Modern discipline is founded on society's becoming increasingly efficient in han-dling relationships of communication, production, and power (216–19). But the excessive and dispersive dynamics of discourse and power relations make them incompatible with the will to truth that dominates this biopower regime. No totalization is achieved, nor do technologies of individuation exhaust the individual. Furthermore, the will to truth, because it is a mode of power, necessarily disrupts its own goals: the more it seeks to localize, identify, and delineate standards of normality and individual forms of delin-quency, the more its own operations of dispersion and discontinuity hinder these aims. Disciplinary and normalizing society is therefore not a society where all individuals become the same, but one where increasing numbers of people are open to discipline for failing to fit ever stricter norms. This incompatibility between aims and results drives the intensification of the will to truth and its expansion into new areas of life.

Adorno perhaps best describes the moral and social malaise of this nihil-istic compulsion. Just as the Christian will to truth, according to Nietzsche, kills God by refusing to allow itself the lie that he exists,[12] so the bourgeois economy destroys the bourgeois morality that formerly nourished it. Moral-ity, for Adorno, was always an ideology hiding domination. Now the lie of

---

12. "You see what it was that really triumphed over the Christian god: Christian morality itself, the concept of truthfulness that was understood ever more rigorously, the father con-fessor's refinement of the Christian conscience, translated and sublimated into a scientific conscience, into intellectual cleanliness at any price. Looking at nature as if it were proof of the goodness and governance of a god; interpreting history in honor of some divine reason, as a continual testimony of a moral world order and ultimate moral purposes; interpreting one's own experiences as pious people have long enough interpreted theirs, as if everything were providential, a hint, designed and ordained for the sake of the salvation of the soul—that is all over now, that has man's conscience against it, that is considered indecent and dishonest by every more refined conscience" (Nietzsche 1974, §357).

morality is admitted to be a lie, or at least it is no longer held as pristine truth. The lie continues, despite being obvious to all, but its status changes: it persists in structuring life, but what now becomes offensive is the failure to lie well.

> The immorality of lying does not consist in the offence against sacrosanct truth. An appeal to truth is scarcely a prerogative of a society which dragoons its members to own up the better to hunt them down. It ill befits universal untruth to insist on particular truth, while immediately converting it into its opposite. Nevertheless, there is something repellent about a lie, and awareness of this, though inculcated by the traditional whip, yet throws light on the gaolers. Error lies in excessive honesty. A man who lies is ashamed, for each lie teaches him the degradation of a world which, forcing him to lie in order to live, promptly sings the praises of loyalty and truthfulness. This shame undermines the lying of more subtly organized natures. They do it badly, which alone really makes the lie a moral offence against the other. It implies his stupidity, and so serves to express contempt. (Adorno 1978, §9)

Capitalist commodity exchange, operating by the equivalence or equality of qualitative differences, corresponds to the reduction of difference to equality in idealist dialectics. Hegel, the exemplary bourgeois thinker, mediates the universal and particular only by replacing the particular or individual with its abstract concept, reducing it to a difference compatible with identity.[13] The result is that Hegel assigns "to individuation, however much he may designate it a driving moment in the process, an inferior status in the construction of the whole. . . . Nowhere in his work is the primacy of the whole

---

13. "Hegel . . . exploited the fact that the nonidentical on its part can be defined only as a concept. To him it was thereby removed from dialectics and bought to identity: the ontical was ontologized" (Adorno 1995, 119). Also: "The concept of the particular is always its negation at the same time; it cuts short what the particular is and what nonetheless cannot be directly named, and it replaces this with identity. . . . For the particular he [Hegel] substitutes the general concept of particularization pure and simple" (173); "there is only one way for Hegelian logic to succinctly identify a universal and an undefined particular, to equate cognition with the fact that the two poles are mediated; and that is for logic . . . not to deal with the particular as a particular at all. His logic deals only with particularity, which is already conceptual" (328); "but if the mediation of the universal by the particular and of the particular by the universal is reduced to the abstract normal form of mediation as such, the particular has to pay the price, down to its authoritarian dismissal in the material parts of the Hegelian system" (329).

doubted" (Adorno 1978, dedication). Similarly, capitalist commodification, extending to all areas of life, crushes the individual through objectification, but does not, for all that, manage to create an ordered whole. "The whole is the false" (§29), but, like other delegitimated values, the ideal of unity persists in a society that can be appraised only negatively as a broken whole.[14] All this time, however, the ideal of individualism is also celebrated.

As commodification enters private life, personal relations are tied to economic advantage, the right friends being a key to success: "indeed the entire private domain is being engulfed by a mysterious activity that bears all the features of commercial life without there being actually any business to transact" (§3). In this environment, where moral conventions have been discredited, a parody of morality continues in the form of tact, which purports to respect the humanity of others but is really dehumanizing: "The question as to someone's health, no longer required and expected by upbringing, becomes inquisitive or injurious, silence on sensitive subjects empty indifference. . . . In the end emancipated, purely individual tact becomes mere lying" (§16). It is not that all intimate relationships and friendships really degenerate into mere show; it is rather that it is never particularly surprising when they reveal themselves as such. Leisure is similarly commodified, planned by the culture industry in ways that coordinate it with, but also liken it to, working time (§84), with consumers being fed as passive recipients. Movies repeat the same tired plots (§94), while the capacities of mass production, having developed independently of what has been produced, use mass advertising to encourage consumption of often unnecessary goods (§§77–78, 129–31).[15] In the end, consumer choice is always false, as public rejection of one product leads only to its replacement by another that is essentially the same: "The culture industry is geared to mimetic regression, to the manipulation of repressed impulses to copy" (§129). It would certainly be hyperbole to say this is the case with every consumer good, but the exceptions ultimately prove the general trend: the mediocrity produced by the marketplace is clear even while it is celebrated as the pinnacle of novelty and freedom.

14. "Measured by its concept, the individual has indeed become as null and void as Hegel's philosophy anticipated: seen *sub specie individuationis*, however, absolute contingency, permitted to persist as a seemingly abnormal state, is itself the essential. The world is systematized horror, but therefore it is to do the world too much honour to think of it entirely as a system; for its unifying principle is division, and it reconciles by asserting unimpaired the irreconcilability of the general and the particular" (Adorno 1978, §72).

15. These critiques of the culture industry are developed most extensively in Adorno and Horkheimer (1997, 120–67).

In public life and work, economic rationality produces a strange mix of hierarchy, fluidity, uncertainty, and compulsions toward conformity. Economic hierarchy becomes more rigid, but social mobility and class membership fluctuate, largely because economic development requires management and technical positions that can be open to everyone because they do not require genuine expertise. At the same time, the insecurity pervading all positions in the workplace becomes "an egalitarian threat" (§124). Even while the disparity between rich and poor escalates, their psychologies converge, since the entire system operates through self-interestedness, domination over others, and the delegation of powers of domination (§§117, 120). The supremacy of technical-minded instrumental rationality and the planned nature of work and leisure blunt critical and reflective thinking. In an objectified world of facts, thinking merely schematizes and reiterates, leaving crucial subjective faculties of reflection and imagination depreciated (§§42, 79, 82). But commodification also turns critique into conformity by submitting intellectual work to the marketplace. Critical distance is impossible because independence from social pressures requires the economic independence of the bourgeois (§6); because intellectual work, no matter how radical, is quickly boxed into abstract categories; and because "all cultural products, even non-conformist ones, have been incorporated into the distribution-mechanisms of large-scale capital," so that "a product that does not bear the imprimatur of mass-production can scarcely reach a reader, viewer, listener at all" (§132). Critical work and political struggle must either be sold or become isolated, compelling a "frantic optimism" (§73) in workers' movements, the manufacture of sham struggles by intellectuals who need theoretical opponents to hype their work (§87), and the dropping of intellectual standards in order to communicate to the masses (§8). In both intellectual and political struggles, a friend/enemy mentality and enforced solidarity dominate (§31). Once again, the exceptions to these trends prove the rule, as the overall moral and intellectual decline leads to the most nihilistic politics and crimes.

A society dominated by the culture industry craves novelty, but in a conformist world this takes the form of a desire for the shocking and sensational. This is a society of decadence in Nietzsche's sense of needing "stronger and stronger and more and more frequent stimulants, such as every exhausted nature is acquainted with" (Nietzsche 1990, 58). It is also, as a totalizing society, intolerant of concrete difference. The familiar liberal "argument of tolerance, that all people and all races are equal, is a boomerang" (Adorno 1978, §66); Marxist attacks on reification and alienation miss the real danger of intolerance of the nonidentical;[16] fascist insanity, although antiliberal, is nevertheless a product of the same forces of industrial capitalism and imperialist competition (§69). Fascism takes the need for stimulus to its extreme: "Fascism was the absolute sensation. . . . In the Third Reich the abstract horror of news and rumour was enjoyed as the only stimulus sufficient to incite a momentary glow in the weakened sensorium of the masses" (§150). Yet this extreme shock is not the opposite of liberal capitalism but rather the logical conclusion of its production of thoughtlessness, its penchant for friend/enemy binaries, its demand for conformity, and the overall decline of its morality. Hence the stupidity of assuming that life can be rebuilt as before after the war and Holocaust: "The idea that after this war life will continue 'normally' or even that culture might be 'rebuilt' . . . is idiotic" (§33); "all post-Auschwitz culture, including its urgent critique, is garbage" (Adorno 1995, 367). Vengeance against the perpetrators of the Holocaust only perpetuates barbarism. "If, however, the dead are not avenged and mercy is exercised, Fascism will despite everything get away with its victory scot-free, and, having once been shown so easy, will be continued elsewhere" (Adorno 1978, §33). The Hegelian idea that history's violence is redeemed as a step in progress is made defunct by the sense carried by the apocalypse's survivors that it is not worth living in a world that allowed this to happen in the first place.

> Hence it may have been wrong to say that after Auschwitz you could no longer write poems. But it is not wrong to raise the less cultural question whether after Auschwitz you can go on living—especially whether one who escaped by accident, one who by rights should have been killed, may go on living. His mere survival calls for the coldness, the basic principle of bourgeois subjectivity, without which there could have been no Auschwitz; this is the drastic guilt of him who was spared. (Adorno 1995, 362–63)

If there is a way past this nihilistic collapse, it cannot take the form of a political thought or practice building new icons or ideals. If "Auschwitz confirmed the philosopheme of pure identity as death" (362), it is also the destruction of identity. The shock of the Holocaust is the shock of nonidentity: "However void every trace of otherness in it . . . in the breaks that belie

16. "If a man looks upon thingness as radical evil, if he would like to dynamize all entity into pure actuality, he tends to be hostile to otherness, to the alien thing that has lent its name to alienation" (Adorno 1995, 191).

identity, entity is still pervaded by the ever-broken pledges of that otherness"
(404). Total destruction opens life to the meaninglessness produced by a
modern society that treats individuals as replaceable cogs. But while the
subsequent indifference to and detachment from life may seem inhuman,
it reveals something that is ultimately perhaps the most human: a resistance
to all ideology that seeks to establish meaning by submitting life to stable
identities and concepts from on high.

> Thinking men and artists have not infrequently described a sense
> of being not quite there, of not playing along, a feeling as if they
> were not themselves at all, but a kind of spectator. Others often
> find this repulsive. . . . A critique of philosophical personalism indi-
> cates, however, that this attitude toward immediacy, this disavowal
> of every existential posture, has a moment of objective truth that
> goes beyond the appearance of the self-preserving motive. "What
> does it really matter?" is a line we like to associate with bourgeois
> callousness, but it is the line most likely to make the individual
> aware, without dread, of the insignificance of his existence. The
> inhuman part of it, the ability to keep one's distance as a spectator
> and to rise above things, is in the final analysis the human part, the
> very part resisted by its ideologists. (363)

Such resistance speaks to concrete difference: "the absolute, as it hovers
before metaphysics, would be the nonidentical that refuses to emerge until
the compulsion of identity has dissolved" (406). Insofar as "the concept of
sense involves an objectivity beyond all 'making': a sense that is 'made' is
already fictitious" (376), nonidentity serves as a groundless ground of another sense—indeed, once freed from the valuations of identity that define
it as chaos and meaninglessness, it becomes the structure of sense as such.
Affirming this structure does not lead to relativistic subjectivism or abstract
individualism (see 35–37) but rather indicates an individualism and subjectivity that go beyond identity: "Only he . . . who would have used his own
strength, which he owes to identity, to cast off the façade of identity—would
truly be a subject" (277). Instead of identity, there is "a togetherness of
diversity" (150)—a structure of differences corresponding to time understood as unchanging form.

The only possibility for thought after Auschwitz is a thinking that thinks
against itself, paying homage to what escapes its concepts without reducing
this excess to a Hegelian-type opposition to be mediated. The task is "to

bring the intentionless within the realm of concepts: the obligation to think at the same time dialectically and undialectically" (Adorno 1978, §98). In this respect, Adorno follows Nietzsche in contending that a certain type of art can combat the will to truth. As Nietzsche maintains, "art, in which precisely the *lie* is sanctified and the *will to deception* has a good conscience, is much more fundamentally opposed to the ascetic ideal than is science" (1967, 3.25). The transgressive power of art, Adorno argues, goes beyond the critical negativity that modern capitalism quickly labels and absorbs. The psychoanalytical account of art as a sublimation of repressed instincts is therefore inadequate: "Artists do not sublimate. That they neither satisfy nor repress their desires, but transform them into socially desirable achievements, their works, is a psycho-analytical illusion" (Adorno 1978, §136). As art embodies "an instinctual impulse expressed uncensored," it "cannot be called repressed even though it no longer wishes to reach the goal it cannot find" (§136). It therefore refuses to accept things as they are, and as such it is bound up with culture without being an achievement of culture. Art, of course, faces the strictures of commodification, which press it to display purpose and value and to communicate a message. Moreover, while great art protests against reality, it seems indistinguishable from low art and kitsch, which are purposeless not because they protest against purpose but because they are feeble (§145). Ultimately, art is caught in irresolvable contradiction: its "aim" is purposelessness, but in being made it is akin to production and so always still faces questions about its function. To hold together such contradictions in their irreducibility and irresolvability, however, is precisely the task of thought, which remains in its own paradox of requiring at once distance from and immersion in the world it challenges (§247). What thinking and art share is that they can sustain themselves only in self-betrayal: "Common to art and philosophy is not the form, not the forming process, but a mode of conduct that forbids pseudomorphosis. Both keep faith with their own substance through their opposites" (Adorno 1995, 15). Their common mode and common task give art and thinking their political and ethical import.

15

Discipline and Normalization

FOUCAULT IS OFTEN CREDITED with the view that power fixes and imposes identities on its subjects, while resistance, which may be considered a form of power, opposes this first power and thereby dissolves or deconstructs power's identity formations.[1] This use of identity as a central term around which power and resistance operate may seem compatible with Foucault's main criticisms of traditional juridico-discursive models of power: that these models fail to appreciate power's positive and productive nature, its differing micro- and macrolevels of operation, and its most complex strategies, which go far beyond those of law and restriction. Yet such readings ulti-

1. Many of these interpretations both ascribe to Foucault an opposition of power and resistance and affirm such an opposition while criticizing Foucault for failing to sustain it. Examples are found in certain normatively oriented theories, variants of which are advanced by Habermas and some feminist political theorists, which seek to justify opposition to dominating forms of power. Here power is seen to fix, repress, and dominate, and Foucault is, in some cases, applauded for demonstrating that certain identities often taken as pregiven or natural—such as those relating to the female body—are actually constituted by this power. Nevertheless, according to these readings, Foucault's thesis of omnipresent power disciplining subjects into socially constituted identities precludes the possibility of any resistance that could oppose power, either because it provides no space for this resistance or because it provides no foundation for a critique of power as domination and thereby falls into relativism. Even Foucault supposedly found this analytic inadequate, which explains his apparent reversal between his writings on power and his writings on the self. Interpretations that follow this pattern include Habermas (1987, 266–92), Fraser (1989), Hartsock (1990), and MacNay (1992).

Another reading, given by certain poststructuralist thinkers such as Judith Butler, supports the view that no simple outside to power exists but makes space for opposing resistance by denying that Foucauldian power can be restricted to domination and repression. It thereby affirms that, for Foucault, resistance is internal to power, but it still accuses Foucault of failing to sustain this view and of ultimately seeking to anchor resistance somewhere outside of power—for example, in bodies and pleasures. See Butler (1987, 217–29; 1990, 93–106). Some readings that take this critical view hold that Foucault fails to locate the site of resistance in a Lacanian-style structural Lack residing within power relations. See, for example, Newman (2001, chapter 4).

mately link the analytic of power to the very will to truth Foucault contests by making identity central to the truths and meanings constructed by power. At times Foucault seems to encourage this oppositional understanding. In "The Subject and Power," for example, he holds that a new analysis of power relations would "tak[e] the forms of resistance against different forms of power as a starting point," and maintains that these resistances "attack everything which . . . forces the individual back on himself and ties him to his own identity in a constraining way" (Foucault 1982, 211–12). But Foucault immediately qualifies this apparent opposition of power and resistance, adding that these resistances "are an opposition to the effects of power which are linked with knowledge, competence, and qualification. . . . What is questioned is the way in which knowledge circulates and functions, its relations to power. In short, the *régime du savoir*" (212). What contemporary forms of resistance challenge, therefore, are not power relations as such, but a knowledge regime that ties the individual in a constraining way to his or her identity. The importance of this subtle shift from power to a power/knowledge regime becomes clear when Foucault examines the efficacy of power relations and denies that this process of fixing is ever really power's aim or effect: modern society, he argues, is *disciplinary*—a fact that, of course, requires a knowledge regime that delineates standards of normality and deviancy against which individuals are subjected—but it is not *disciplined*.

> What is to be understood by the disciplining of societies in Europe since the eighteenth century is not, of course, that the individuals who are part of them become more and more obedient, nor that they set about assembling in barracks, schools, or prisons; rather that an increasingly better invigilated process of adjustment has been sought after—more and more rational and economic— between productive activities, resources of communication, and the play of power relations. (219)

If power really interpellated[2] or compelled individuals toward a set of normal identities, then the irony of both *Discipline and Punish* and *The History of Sexuality,* volume 1, is that no "normal" individual ever appears.

2. On the Althusserian notion of interpellation, see Althusser (1984); also Butler (1993, 121–24).

From a perspective governed by the will to truth, this amounts to a failure of power to discipline and normalize. But Foucault suggests that this was never actually the goal. Given that an examination of the various criminal delinquents that modern society seeks to police reveals that they have always already passed through a myriad of institutions supposedly designed to correct them,[3] how could it ever be said that power aimed to normalize individuals in the first place? Nor can one say that power forms identifiable deviants, since these individuals persistently escape classification: society's Others, such as Herculine Barbin and Pierre Rivière, remain enigmatic despite all the energies of medical and legal discourse to individuate, rank, and comprehend them (see Foucault 1975 and 1980a). We must therefore understand the aim of disciplinary and normalizing power relations, which, like all power relations, "are both intentional and nonsubjective" (Foucault 1990a, 94), to be to produce forms of otherness that, while not necessarily knowable, are compatible with modern liberal capitalism and can be invigilated in a drive to increase social and economic efficiency—even if this project has numerous built-in inefficiencies. The prison, for example, "has succeeded extremely well in producing delinquency, a specific type, a politically or economically less dangerous—and, on occasion, usable—form of illegality" (Foucault 1979, 277). In other words, it has produced manageable forms of difference. No successfully constituted identity, whether normal or deviant, is requisite for this objective.

Historically, the rise of disciplinary and normalizing power strategies correlates with the decline of hierarchical and centralized systems of sovereign power, the development of capitalism and free market economies, and the rise of social contract understandings of government. A new form of governmentality is required, and it cannot rely on old threats of punishment and death from the sovereign but must instead promote the life and efficiency of the body politic. Individuals must be constituted in certain ways to counterbalance the new economic and social freedoms that come with these changes.[4] This does not mean that individuals must become "the

3. "The delinquent is an institutional product. It is no use being surprised, therefore, that in a considerable proportion of cases the biography of convicts passes through all these mechanisms and establishments, whose purpose, it is widely believed, is to lead away from prison" (Foucault 1979, 301).

4. "It is democracy—or better still, the liberalism that matured in the nineteenth century—which has developed extremely coercive techniques that in a certain sense have become the counterbalance to a determinate economic and social 'freedom.' Individuals certainly could not be 'liberated' without educating them in a certain way" (Foucault 1991, 171).

same" as each other or must be reduced to "docility,"[5] but it does require that they be constituted against a norm that measures them. The norm and the various deviant positions mapped out in opposition to it act as markers to coordinate the deployment and operation of this new power's various techniques—observation, testing, classification, and confession—which operate through strict control of space and (chronological) time.[6] Even if these strategies never effectively compel people toward the norm, this only justifies intensification of the disciplinary regime. A simulacrum of binary categories thereby overlays the network of disciplinary power relations, the former seeming to give the latter meaning and sense even while the products of these power relations elude the grasp of these categories. Unlike the techniques of the earlier sovereign form of power, these strategies operate through both punishment and reward, policing nonconformity rather than disobedience. They are diffused throughout society, and while they involve relations of authority and subordination, they crucially require the subordinate's participation in his or her disciplining or subjection. But to be subjected to an identity or corrected on the basis of a normalizing judgment is not to be interpellated into this identity. The pettiness and banality of the disciplining—making the child stand in the corner or the prisoner eat dry bread with water (Foucault 1977a, 210), with similarly puerile rewards given for "good behavior"—ultimately do little to encourage conformity, although this does nothing to prevent their regular and frequent use.

Disciplinary and normalizing strategies also differ from those of sovereign power in operating primarily at the microscopic level. It is not that the microscopic was absent in premodern times, but the old power's most prominent mechanisms did not make this level the explicit site of their interventions. This makes the juridico-discursive models of power, which

5. The common charge that Foucault sees an all-encompassing modern power reducing individuals to "docile bodies" fails to recognize that this is simply the ideal of a military model of society that arose in the classical age: "Historians of ideas usually attribute the dream of a perfect society to the philosophers and jurists of the eighteenth century; but there was also a military dream of society; its fundamental reference was not to the state of nature, but to the meticulously subordinated cogs of a machine, not to the primal social contract, but to permanent coercions, not to fundamental rights, but to indefinitely progressive forms of training, not to the general will but to automatic docility" (Foucault 1979, 169). Undoubtedly the mechanisms of disciplinary society could not have arisen without such an ideal, but this hardly means they have had the desired effects.

6. Crucially, while discipline requires control over space, this is not necessarily a closed space: "Discipline sometimes requires *enclosure*, the specification of a place heterogeneous to all others and closed in upon itself" (Foucault 1979, 141). This point should answer interpretations such as Hardt and Negri's (2000, 24, 330), which holds that Foucault's disciplinary society has been superseded by a more decentralized and open form of "control society."

treat power as a kind of "anti-energy" (Foucault 1990a, 85) working through binary categories of "licit and illicit, permitted and forbidden" (83), insufficient to understand a modern society "that has been more imaginative, probably, than any other in creating devious and supple mechanisms of power" (86) and that "has gradually been penetrated by quite new mechanisms of power that are probably irreducible to the representation of law" (89). This criticism also applies to psychoanalytic models that hold the law to constitute the very desire it prohibits or restricts: the question of whether desire is prior to the law or the law's product "is beside the point" (89). Historically, then, there arises a new focus on a certain domain of power relations, rather than, as some interpreters seem to suggest, the invention of a new, microscopic level of power.

Microscopic and macroscopic are neither simply external to one another nor internal and identical.[7] They are immanent to each other and reciprocally determining. Microscopic relations are constitutive but hidden. They operate on the plane where individuals are constituted and disciplinary practices are microscopic because they function primarily at this constitutive level. Macroscopic power relations hold between precariously constituted subjects and on this level power can be understood as a possession of one individual who chooses to exercise or not exercise it against another. Power could not operate without the delineation of subject positions that allow the agents associated with them to possess powers with which they can dominate others, but this situation is a consequence of the microscopic flows that condition the appearance of these positions and their relations to one another. Subject positions and relations being neither natural nor pregiven, they are necessarily products of power. But this is not a power possessed by agents. Instead, micropower relations form "a network of relations, constantly in tension, in activity" (Foucault 1979, 26). They are "the moving substrate of force relations which, by virtue of their inequality, constantly engender states of power, but the latter are always local and unstable," while the macroscopic power relations they establish are "the over-all effect that emerges from all these mobilities, the concatenation that rests on each of them and seeks in turn to arrest their movement" (Foucault 1990a, 93). But through the "double conditioning" between them (see 99–100), the two domains both reinforce and undermine each other. A police officer on the

---

7. "There is no discontinuity between them, as if one were dealing with two different levels (one microscopic and the other macroscopic); but neither is there homogeneity (as if the one were only the enlarged projection or the miniaturization of the other)" (Foucault 1990a, 99–100).

street, a judge in the courtroom, a teacher in school, all have powers that they can choose to exercise against others, but only by virtue of a context of meanings and subject positions that they share with those subordinate to them and through a sense of discipline that pervades the entire situation. Moreover, the authority of superiors is always unstable, open to actions by subordinates that subvert their legitimacy, such actions often using rules and strategies made available by the same power relations that constitute the hierarchies.[8] At the same time, superiors may use their possessed pow- ers to reinforce the microscopic hierarchies supporting them, but these uses are often ambiguous. The use of police force, for example, can both rein- force and undermine police authority—indeed, often it does both at once.

If the microscopic level generates the appearance and positioning of sub- jects and objects, then it is the dynamic of discursive formations that ex- plains how power at this level operates. Following this dynamic, subjects and objects are formed in intersections that link heterogeneous domains and are replete with divergences and discontinuities. Micropower relations are thus relations of disjunction, which arise with and permeate the prac- tices they exceed: they "are not in a position of exteriority with respect to other types of relationships (economic processes, knowledge relationships, sexual relations), but are immanent in the latter; they are the immediate effects of the divisions, inequalities, and disequilibriums which occur in the latter, and conversely they are the internal conditions of these differentia- tions" (Foucault 1990a, 94). This accounts for an important element of resistance residing within power relations, which follows from the way power, as a synthesis of disjunction, is always a relation of discontinuity and disequilibrium. On the one hand, the crisscrossing discursive domains formed by disjunction remain porous and open to change; on the other hand, any subject, object, or corresponding knowledge carries ambivalences due to the heterogeneity and flux of its necessary constituents. The nine- teenth-century discourses on human sexuality, mentioned before in relation to the will to truth, provide one example. Another appears in the way diver-

8. In the United States, an authority figure's authority might be problematized by defi- antly asking, "who do you think you are?" In societies that are more status oriented, the question might be, "do you know who you are talking to?"—a generally ineffective strategy in the United States that has been known to get celebrities arrested on the spot. These tactics are dictated by the rules of discursive games, and so they are not outside or opposed to power but rather result from the discontinuities that power relations inevitably establish. They pro- vide possibilities to disrupt, even if only briefly, microscopic hierarchies. Even if such strate- gies are not revolutionary—and one can end up getting arrested anyway—it would be hard to suggest that they are not political.

gent oppositional categories reinforce the uncertain status of various identities deemed to be pathological: "When a judgment cannot be framed in terms of good and evil, it is stated in terms of normal and abnormal. And when it is necessary to justify this last distinction, it is done in terms of what is good or bad for the individual" (Foucault 1977b, 230). In such instances, the need to shift from one opposition to another both sustains the condemnable status of the identity in question and problematizes its position.

Resistance sometimes takes an oppositional stance against power—for example, when the treatment of prisoners produces resentment toward the system and ultimately leads to recidivism (see Foucault 1979, 264–68)—and there are, occasionally, even "great radical ruptures, massive binary divisions" (Foucault 1990a, 96). But the complex discontinuities of micropower relations and their imbrication in macroscopic relations reveal a range of resistances beyond mere opposition. The heterogeneities within all levels of power relations produce incompatibilities within and among modern disciplinary institutions and practices. Resistances thus take various roles as "adversary, target, support, or handle in power relations" (95), while being "the odd term in relations of power" (96).[9] Because authoritative subjects are constituted in one domain of convergence while the objects form elsewhere, cross-purposes and frictions arise between institutions that can function only by acting together. The family and the psychiatric profession, for example, cooperate to monitor sexuality in the home, but also clash when psychiatrists seek to institutionalize family members (see 111–12). But the dispersive nature of discursive power, which overflows all categories of knowledge, also produces the indeterminate mutations that make it excessive and dangerous. Thus the exercise of power to observe and investigate sexual deviance, by virtue of its inherent voyeurism and the erotic "hide-and-seek" games it encourages, ends up proliferating rather than controlling sexual desire (44–45).

Despite superficial appearances, power operates only through indeterminate dispersion, even while it gives rise to a semblance of stable identities and oppositions and to a will to truth that, seeking to secure identity, is

---

9. Elsewhere, Foucault acknowledges that both dispersed procedures of power and multiple forms of resistance can be integrated into "global strategies," but he still warns: "one should not assume a massive and primal condition of domination, a binary structure with 'dominators' on one side and 'dominated' on the other, but rather a multiform production of relations of domination which are partially susceptible of integration into overall strategies" (Foucault 1980b, 142).

driven to locate, observe, measure, and know all differences through binary categories of normality and deviance. In the end, however, persistent resistance does little to undermine the disciplinary regime, and the entire system remains trapped in its own logic, unable to do anything more than increase its policing. The seeming failure of discipline only reinforces its apparent necessity—"for the past 150 years the proclamation of the failure of the prison has always been accompanied by its maintenance" (Foucault 1979, 272)—and results in its replication throughout all areas of life: "Is it surprising that prisons resemble factories, schools, barracks, hospitals, which all resemble prisons?" (228). Disciplinary and normalizing powers produce neither normal citizens nor deviants but enigmas, which is sufficient to sustain the disciplinary structure necessary for modern governmentality.

16

Time, Guilt, and Overcoming

ONE DOES NOT NEED FREUD to understand how trauma initiates a repetition operating beyond any reference to pleasure or the pleasure principle. The death of a loved one, the breakdown of a marriage, or even a minor public embarrassment—regardless of whether these occur quickly or develop gradually in chronological time—easily result in a compulsion to replay in memory and behavior the events and the context that produced them. Freud asserts that through repetition the trauma of the event can be mastered, allowing the psychic system to repair itself and the pleasure principle to return to dominance.[1] Yet nothing in bare repetition compels this result. Moreover, even while none of these repetitions are identical, just as no two grains of sand or drops of rain are the same, this difference is not enough to move beyond the event and create something new. Is it any wonder why Deleuze says that difference without concept—the indifferent difference that distinguishes otherwise identical repetitions without affecting their fundamental identity—is not the essence of repetition?[2] Everyone has expe-

---

1. "We may assume, rather, that dreams are here helping to carry out another task, which must be accomplished before the dominance of the pleasure principle can even begin. These dreams are endeavouring to master the stimulus retrospectively, by developing the anxiety whose omission was the cause of the traumatic neurosis. They thus afford us a view of a function of the mental apparatus which, though it does not contradict the pleasure principle, is nevertheless independent of it and seems to be more primitive than the purpose of gaining pleasure and avoiding unpleasure" (Freud 1957a, 32).

2. "When we define repetition as difference without concept, we are drawn to conclude that only extrinsic difference is involved in repetition; we consider, therefore, that any internal 'novelty' is sufficient to remove us from repetition proper and can be reconciled only with an approximative repetition, so-called by analogy. Nothing of the sort is true. For we do not yet know what is the essence of repetition, what is positively denoted by the expression 'difference without concept,' or the nature of the interiority it may imply. Conversely, when we define difference as conceptual difference, we believe we have done enough to specify the concept of difference as such. Nevertheless, here again we have no idea of difference, no concept of difference as such. . . . We therefore find ourselves confronted by two questions:

rienced the rut of being absorbed in such repetitions and their meaningless differences. This malaise would be inescapable—is that not genuine nihilism?—if there were not something literally coming from the "outside" that could compel change. "God" is one answer to the question of what this outside force might be. But another answer is time—or, rather, that dimension of time's structure that conditions the movement or change of entities "in time." This dimension, contra Bergson, cannot be linked to the past and memory: the past is not the "outside" that orients time and things in time.

Heidegger's *Being and Time* proposes an existential analytic of Dasein's Being-in-the-world, which, worked out in temporal terms, lays the groundwork for addressing the question of the meaning of Being and its temporal horizon. Temporality does not mean "being in time" but refers instead to the ecstatic structure of Dasein's Being (see Heidegger 1962, 39–40, 375) and ultimately to the structure of Being as such. When Dasein's structure as care—the form of Being-thrown-ahead-of-itself and Being-alongside-entities-encountered in-the-world,[3] which accounts for Dasein's relations to the ready-to-hand and the present-at-hand, its Being-with-others, its states of mind, moods, and understandings, and its falling and absorption into the world of the "they"—is recast in temporal determinations, it becomes clear that "the future has a priority in the ecstatical unity of primordial and authentic temporality" (378). Anxiety, the state of mind arising from the uncanniness experienced when the world loses its significance (393), indicates authentic Being-towards-death to be Dasein's fundamental comportment toward its utmost potentiality-for-Being. The event of death, which is the possibility of radical nonexistence (307), is futural, not in the sense of being "not yet" in time (373) but as something ever present yet seemingly coming from nowhere (231). It is possible because "Dasein, *as being*, is always coming towards itself—that is to say, in so far as it is futural in its Being in general" (373). Dasein's Being-towards-death means that its structure is "whole" by virtue of including this nullity within it. In Being-towards this absolute limit, Dasein's Being becomes an issue for it. It is the future—albeit not the future of linear, chronological time—that orients the dynamic of Dasein's Being.

---

what is the concept of difference—one which is not reducible to simple conceptual difference but demands its own Idea, its own singularity at the level of Ideas? On the other hand, what is the essence of repetition—one which is not reducible to difference without concept, and cannot be confused with the apparent character of objects represented by the same concept; but bears witness to singularity as a power of Ideas?" (Deleuze 1994, 27).

   3. See Heidegger (1962, 237) for the first definition of the meaning of "care."

In its inauthentic mode, Dasein flees from itself and its death into the public idle talk of the "they," which interprets death as belonging to someone else or to no one in particular (Heidegger 1962, 297). It falls into concernful Being-in-the-world with meaningful and calculable entities that are ready-to-hand and present-at-hand, reckoning with time in an ordinary way—although even here the vulgar conception of time as a sequence of "nows" is insufficient (467). In its authentic mode of anticipatory resoluteness, however, Dasein faces death and authentically chooses from among the possibilities given to it in its Situation. Thrown toward a future that approaches it, "Dasein understands itself with regard to its potentiality-for-Being, and it does so in such a manner that it will go right under the eyes of Death in order thus to take over in its thrownness that entity which it is itself, and to take it over wholly" (434). The possibilities that resolute Dasein projects and chooses, however, "are not to be gathered from death" (434). Instead, they are disclosed "*in terms of the heritage* which that resoluteness, as thrown, *takes over*" (435). This heritage, which is a fate or destiny wrapped up with Dasein's solicitous Being-with-others (435–36), exists because Dasein, in being thrown, carries its past with it: "Only because care is based on the character of 'having been,' can Dasein exist as the thrown entity which it is. 'As long as' Dasein factically exists, it is never past [vergangen], but it always is indeed as already having *been*, in the sense of the 'I *am*-as-having-been'" (376). Through authentic choice in the "moment of vision," Dasein determines not only its present and future, but also its past, so that "taking over thrownness . . . is possible only in such a way that the futural Dasein can *be* its ownmost 'as-it-already-was'—that is to say, its 'been'" (373). This choice effects a genuine repetition, which brings on the new.

> The repeating of that which is possible does not bring again [Wiederbringen] something that is "past," nor does it bind the "Present" back to that which has already been "outstripped." Arising, as it does, from a resolute projection of oneself, repetition does not let itself be persuaded of something by what is "past," just in order that this, as something which was formerly actual, may recur. Rather, the repetition makes a *reciprocative rejoinder* to the possibility of that existence which has-been-there. But when such a rejoinder is made to this possibility in a resolution, it is made *in a moment of vision; and as such* it is at the same time a *disavowal* of that which in the "today," is working itself out as the "past." Repetition does not abandon itself to that which is past, nor does it aim

at progress. In the moment of vision authentic existence is indifferent to both these alternatives. (437–38)

Fate and repetition thereby constitute Dasein's authentic historicality (442), allowing Dasein to become what it is.

Authentic Being-towards-death is also "freedom towards death," which releases Dasein from the illusions of the "they" (311). It is an ontological possibility whose ontic correlate is the call of conscience. The inability to prove or disprove the reality of conscience merely indicates that it has a different ontological status from the ready-to-hand and the present-at-hand (314). Its call "comes *from* me and yet *from beyond me*" (320) and so is not willed by the self, by another self existing alongside, or by an alien but present-at-hand power such as God. It is essentially a call from the "outside," issuing from Dasein's primordial structure, from authentic Dasein finding itself in the depths of uncanniness (321–22). In returning Dasein to itself, conscience speaks in terms of guilt: "All experiences and interpretations of the conscience are at one in that they make the 'voice' of conscience speak somehow of 'guilt'" (325). Conceived primordially, guilt does not refer to ontic debt—and this includes theological conceptions of guilty indebtedness—which is a derivative and inauthentic form (see 326–29). Rather, it speaks to the nullity at the center of Dasein's structure, residing there by virtue of Dasein's Being-thrown and corresponding to its burden of responsibility for Being-free: "*Care itself, in its very essence, is permeated with nullity through and through.* Thus 'care'—Dasein's Being—means, as thrown projection, Being-the-basis of a nullity (and this Being-the-basis is itself null). This means that *Dasein as such is guilty,* if our formally existential definition of 'guilt' as 'Being-the-basis of a nullity' is indeed correct" (331). This nullity is not a lack or corruption of some prior state of fullness and purity; rather, it has a positive content as the condition for Dasein to be morally good or corrupt and amounts, in this sense, to Dasein's free will. It speaks to Dasein's original responsibility for itself, which is consistently obscured by inauthentic modes of existence. In coming face to face with and responding authentically to this responsibility, Dasein "has chosen itself" (334).

Heidegger, of course, was ultimately unsatisfied with his formulation of the problem of Being in *Being and Time*. Approaching Being through Dasein's Being-in-the-world risks reinforcing the very subjectivity the study seeks to surpass because it cannot easily progress to the more primordial relationship of man to Being itself (see Heidegger 1993, 231–32). The analyses in *Being and Time* thus amount to "half attempts" (Heidegger 1972, 44) to work out the question of Being, efforts that were burdened by the post-Kantian framework of the phenomenological context in which it was written. They articulate, against the "vulgar" conception of time, the various ways that past and future, rather than being "not yet" and "no longer existing" instants extending from a present "now," both have presence in Dasein's three-dimensional temporality. But this temporality and its temporalization presuppose another facet of time that allows these various dimensions to become present.[4]

> In the approaching of what is no longer present and even in the present itself, there always plays a kind of approach and bringing about, that is, a kind of presencing. We cannot attribute the presencing to be thus thought to one of the three dimensions of time, to the present, which would seem obvious. Rather, the unity of time's three dimensions consists in the interplay of each toward each. This interplay proves to be the true extending, playing in the very heart of time, the fourth dimension, so to speak—not only so to speak, but in the nature of the matter.
> True time is four-dimensional. (15)

Being is this "giving presence" or presencing; man, as the being essentially able to respond to Being, relates to Being reciprocally and constitutively in a "belonging together," whereby their difference remains irreducible and discordant;[5] and time is the "event of appropriation," which relates man and Being through their difference: "The event of appropriation is that realm, vibrating within itself, through which man and Being reach each other in their nature, achieve their active nature by losing those qualities with which

---

4. This corresponds to Deleuze's (1989, 155) distinction between the order of time and the series of time, noted briefly in this work's fourth reflection.

5. "If we think of belonging *together* in the customary way, the meaning of belonging is determined by the word together, that is, by its unity. In that case, 'to belong' means as much as: to be assigned and placed into the order of a 'together,' established in the unity of a manifold, combined into the unity of a system, mediated by the unifying center of an authoritative synthesis. . . . However, belonging together can also be thought of as *belonging* together. This means: the 'together' is now determined by the belonging. Of course, we must still ask here what 'belong' means in that case, and how its peculiar 'together' is determined only in its terms. . . . Enough for now that this reference makes us note the possibility of no longer representing belonging in terms of the unity of the together, but rather of experiencing this together in terms of belonging" (Heidegger 1969, 29). Also: "That relation is revealed as discordant. The question still remains whether the discordancy of our relation to Being lies in us or in Being itself; the answer to that question may once again decide something important about the essence of that relation" (Heidegger 1979–87, 4:194).

metaphysics has endowed them" (Heidegger 1969, 37). By constituting both presencing and the being that can respond to presencing, the event of appropriation is also the originary fourth dimension of time, giving or donating Being to Dasein's temporality by functioning as its differenciator.

> The dimension which we call the fourth in our count is, in the nature of the matter, the first, that is, the giving that determines all. In future, in past, in the present, that giving brings about to each its own presencing, holds them apart thus opened and so holds them toward one another in the nearness by which the three dimensions remain near one another. For this reason we call the first, original, literally incipient extending in which the unity of true time consists "nearing nearness," "nearhood" (*Nahheit*), an early word still used by Kant. But it brings future, past and present near to one another by distancing them. For it keeps what has been open by denying its advent as present. (Heidegger 1972, 15)

This step back from *Being and Time*, however, seems not to compel any reconsideration of the text's portrayal of Dasein's Being-in-the-world and specifically of the associations it draws between death, choice, conscience, and guilt. On the contrary, Heidegger insists that the analytic of Dasein's authentic Being, despite its inadequacy in navigating a path to Being, was always motivated by the quest to discern man's higher destiny in relation to Being: "Man is the shepherd of Being. It is in this direction alone that *Being and Time* is thinking when ecstatic existence is experienced as 'care'" (Heidegger 1993, 234). But does this settle the matter, or does it simply foreclose consideration of other ethical modes of existence that primordial time might indicate? Heidegger's uncertain relationship with Nietzsche is suggestive, as *Being and Time* makes at least three major claims that invite Nietzschean objections: the declaration that all voices of conscience, good and bad, speak in the language of guilt, against Nietzsche's distinction of bad conscience, which is associated with guilt, from noble conscience, which is not;[6] the assertion that all prior conceptualizations of guilt concern ontic debt only, against Nietzsche's portrayal of guilt and free will as moral-

---

6. This is the conscience of the one whose mastery over himself and his circumstances creates "the proud awareness of the extraordinary privilege of *responsibility,* the consciousness of this rare freedom, this power over oneself and over fate" (Nietzsche 1967, 2.2) and thus gives this being "the right to make promises" (2.1). Crucially, for Nietzsche, this conscience exists prior to "bad conscience," which moralizes debt into the concept of guilt (see 2.4, 20–22).

izations of debt that move them onto the ontological register;[7] and the contention that the possibilities for Dasein's authentic choice come from the heritage it receives from its Volk, against the Nietzschean overman's refusal of such heritage in order to create.[8] Whereas Heidegger links conscience, guilt, and authentic return to an authentic community, Nietzsche separates them. While there remains with Nietzsche a command from conscience to become what one is ("*What does your conscience say?*—'You shall become the person you are'" [Nietzsche 1974, §270]), this occurs through redemptive overcoming rather than the assumption of a burden of guilt. If Nietzsche also offers an ethics suitable for the end of metaphysics, it is surely different from Heidegger's.

In this regard, it is not surprising that when Heidegger confines Nietzsche to the history of metaphysics, he focuses on the issue of time, declaring the eternal return to be a return of the same.[9] Heidegger recognizes that

7. Heidegger claims that "the Being-guilty which belongs primordially to Dasein's state of Being, must be distinguished from the *status corruptionis* as understood in theology. Theology can find in Being-guilty, as existentially defined, an ontological condition for the factical possibility of such a *status*. The guilt which is included in the idea of this *status*, is a factical indebtedness of an utterly peculiar kind. It has its own attestation, which remains closed off in principle from any philosophical experience" (1962, 496n2). Yet Nietzsche's analysis of guilt, which becomes universal only when the image of Christ's sacrifice moralizes debt into something uncreditable, shows precisely that guilty indebtedness is *not* factical: "God himself sacrifices himself for the guilt of mankind, God himself makes payment to himself, God as the only being who can redeem man from what has become unredeemable for man himself—the creditor sacrifices himself for his debtor, out of *love* (can one credit that?), out of love for his debtor!" (1967, 2.21). Moreover, Nietzsche argues, this guilty conscience and its reference to an originary responsibility and free will has its genesis in "that will to self-tormenting, that repressed cruelty of the animal-man made inward and scared back into himself . . . who invented the bad conscience in order to hurt himself after the *more natural* vent for this desire to hurt had been blocked" (2.22). While Heidegger may be correct that an existential condition of Being-free is prior to any notion of guilt as corruption or ontic debt, Nietzsche challenges the conclusion that ontological guilt is anything more than a limited and misguided interpretation of Being-free.

8. "For one should not overlook this fact: the strong are as naturally inclined to *separate* as the weak are to *congregate*; if the former unite together, it is only with the aim of an aggressive collective action and collective satisfaction of their will to power, and with much resistance from the individual conscience; the latter, on the contrary, *enjoy* precisely this coming together—their instinct is just as much satisfied by this as the instinct of the born 'masters' (that is, the solitary, beast-of-prey species of man) is fundamentally irritated and disquieted by organization" (Nietzsche 1967, 3.18).

9. Heidegger consistently renders Nietzsche's doctrine as "the eternal return of the same" (*die ewige Wiederkehr des Gleichen*), even though Nietzsche rarely speaks in such terms. Indeed, only in the first plan, written in August 1881 "in Sils-Maria, 6,000 feet above sea level and much higher above all human things," does Nietzsche speak of a "return of the same" (*die Wiederkunft des Gleichen*; see Heidegger 1979–87, 2:74–75). The second plan Heidegger examines (2:77), from the same time, speaks of an eternal return of all things (*die ewige Wiederkunft aller Dinge*), but all subsequent plans simply call the doctrine "the eternal

Zarathustra's dismissals of his animals' and the dwarfs' renderings of the doctrine mean that it is more than a return of identical events in chronological time (see Heidegger 1979–87, 2:37–62). Nevertheless, Heidegger claims, the genuine doctrine differs only by inserting the overman into the returning circle: while the dwarf and animals interpret the doctrine abstractly from the observer's external standpoint, the overman actively wills a choice from within the constitutive moment of time in order to determine what events will return in time (2:182–83). Grasping the essence of the eternal return thus "requires that one adopt a stance of his own within the 'Moment' itself, that is, in time and its temporality" (2:44). Only from this immanent perspective is it possible to understand Zarathustra's statement that past and future "offend each other face to face" in the Moment (see Nietzsche 1966, "On the Vision and the Riddle," 157–58; Heidegger 1979–87, 2:56). In this way the eternal is defined not as an infinite sequence of "nows" but as a moment of vision that, bending back on itself (Heidegger 1979–87, 1:20), brings together time's three dimensions, past, present, and future, and makes them return. Nevertheless, this moment constitutes only man's Being-in-the-world, and the eternal return concerns only the structure of the world, of being as a whole, not Being itself (2:108–9). Moreover, by maintaining the old metaphysical determinations of being as fixation and becoming as change, the eternal return, which Nietzsche says stamps being onto becoming (Heidegger frequently cites Nietzsche 1968, §617), perpetuates the spirit of revenge by taking vengeance on the transient (Heidegger 1979–87, 2:228–30). Like all metaphysicians, Heidegger argues, Nietzsche sides with being over becoming (2:173) and therefore fails to grasp Being as the event of appropriation. Heidegger can thus easily shift, after his disenchantment with Nietzsche in 1939, from seeing the overman as an authentic chooser to seeing him as the embodiment of metaphysics' final move to master the world through technology (see, for example, 3:174–80), since he can portray Nietzsche as having never asked the grounding question of Being.

However, what if, contra Heidegger, the eternal return is interpreted as an account of Being's temporality and not simply man's Being-in-the-world?

---

return" (*die ewige Wiederkunft*). Moreover, even in the early unpublished notes Nietzsche remains undecided over whether or not a return of the same follows from the finitude of cosmic energy and the infinity of chronological time (see David Farrell Krell's analysis in 2:261–62). Heidegger's use of Nietzsche's notes may be checked against either the *Grossoktavausgabe* used by Heidegger or the *Kritische Gesamtausgabe* of Nietzsche's works (Nietzsche 1967–79), using the references supplied in the appropriate editor's footnotes to Heidegger's Nietzsche lectures.

Moreover, what if, following Deleuze, it is no longer interpreted as a return of the same? Then the eternal return would render a structure comparable to Heidegger's event of appropriation, whereby a second-order difference or differenciator brings together differences and dimensions of time, linking them through their difference rather than through identity. An "outside" element is thereby introduced into the structure of time, one that is neither past, present, nor future: an ultimate Event, which appropriates time's dimensions and which, being unlocalizable and untimely yet ever present in all events occurring "in time," is always "new." The disjunctive synthesis effected by this Event causes difference and differenciation rather than identity to return, making the eternal return an always decentered circle. But the becoming it unfolds in time is only the external aspect of the dissymmetry and dissynchrony within the moment itself. Moreover, there arises with this decentered becoming a simulation of identity by which the differenciator disguises itself. The eternal return is thus a structure in which identity and continuity always float on the surface of dynamic and dispersive becomings of difference.[10]

There is also a conventional circle involved in the return, but its role is preparatory. When Zarathustra is asked to persuade cripples of his teachings on overcoming, he dismisses the idea of removing their burdens or healing their disabilities; instead, he declares his preference for those missing an eye or limb over men of *ressentiment,* those "inverse cripples" who are "nothing but a big eye or a big mouth or a big belly or anything at all that is big" (Nietzsche 1966, "On Redemption," 138).[11] The latter are too attached to their superficial stabilities and identities to be able to overcome past wounds. They cannot let go of their suffering or themselves, their *ressentiment* taking the form of the question: "why did this shit happen to *me?*" The past, Zarathustra notes, is replete with injuries and injustices, which leave only "fragments and limbs and dreadful accidents—but no human beings" (138) and threaten to make life unbearable: "The now and the past

---

10. Heidegger recognizes this aspect of Nietzsche's thought, wherein the metaphysical conceptions of Being as permanence and truth as fixation become mere semblance, appearance, and error (see, for example, Heidegger 1979–87, 1:214–16). Moreover, he acknowledges throughout that Nietzsche does more than simply invert metaphysical thought and that appearance is no longer thought in the same terms as when it was opposed to essence. Nevertheless, this affirmation of semblance for Heidegger merely paves the way for a subjective assertion of mastery: "Nietzsche's metaphysics directly posits untruth in the sense of error as *the* essence of truth. Truth—so qualified and conceived—fashions for the subject an absolute power to enjoin what is true and what is false" (4:145).

11. Heidegger (1979–87, 2:220–28) focuses on this section of *Zarathustra* when arguing that Nietzsche does not escape the spirit of revenge.

on earth—alas, my friends, that is what *I* find most unendurable; and I should not know how to live if I were not also a seer of that which must come" (138–39). This existential fact leads to the first directive of the eternal return: "to recreate all 'it was' into a 'thus I willed it'—that alone should I call redemption" (139). But Zarathustra immediately warns that this redemptive act of will is also a trap: "the will itself is still a prisoner. Willing liberates; but what is it that puts even the liberator himself in fetters? 'It was'—that is the name of the will's gnashing of teeth and most secret melancholy. . . . Alas, every prisoner becomes a fool; and the imprisoned will redeems himself foolishly" (139). Assuming responsibility for the past and thereby choosing itself, the will still remains parasitically attached to the past and to the spirit of revenge, which is nothing other than "the will's ill will against time and its 'it was'" (140). Its act remains consistent with viewing existence as punishment, and "'punishment' is what revenge calls itself; with a hypocritical lie it creates a good conscience for itself" (140). This spirit of revenge ultimately develops into a madness that takes the transience of all existence to indicate that what passes away is condemnable and deserves to pass away. Accepting the burden of past contingencies might reconcile the will with the past, but "that will which is the will to power must will something higher than any reconciliation" (141). Genuine redemption—conscience without guilt—therefore requires that "the creative will says to it, 'But thus I will it; thus shall I will it'" (141). This second move restores innocence to becoming by affirming all existence as worthy of returning—which is not the same as finding transience unacceptable.[12] Indeed, as an act of self-overcoming, willing the eternal return necessarily excludes the will itself from returning, dissolving the "I" or ego. After offering the teaching of eternal return as the path to others' redemption, Zarathustra therefore demands his own redemption through a dying or going under (see Nietzsche 1967, "On Old and New Tablets," 198). The eternal return is inseparable from this "impersonal death" or "going under," which opens the self to multiplicity. Redemption is necessarily an act of transmutation.

When Deleuze speaks of the eternal return as the third synthesis of time, he argues that within it "the present is no more than an actor, an author, an agent destined to be effaced; while the past is no more than a condition

---

12. "Against the value of that which remains eternally the same (*vide* Spinoza's naiveté; Descartes' also), the values of the briefest and most transient, the seductive flash of gold on the belly of the serpent *vita*" (Nietzsche 1968, §577).

operating by default" (1994, 94). Past sufferings provide the default condition for the act of overcoming—the affirmation of eternal return—yet from this initial perspective the act appears impossible, too great for the wounded ego that would have to carry it out. The performance of the act in the present moment thus requires a consolidation of the ego in relation to an ego ideal, which makes the self equal to the task. But while overcoming finds its origin in this unified ego as its author and agent, "the event and the act possess a secret coherence which excludes that of the self; . . . they turn back against the self which has become their equal and smash it to pieces, as though the bearer of the new world were carried away and dispersed by the shock of the multiplicity to which it gives birth: what the self has become equal to is the unequal in itself" (89–90; see also 91–92, 110–11). As a resonance of series through a differenciator, the eternal return effects "a *forced movement* the amplitude of which exceeds that of the basic series themselves" (117). In this way it institutes a creativity, realized in thinking and in the thought of eternal return, that makes possible a break with the past. In the final instance, then, the eternal return concerns only the future, by virtue of its "expelling the agent and the condition in the name of the work or product; making repetition, not that from which one 'draws off' a difference, nor that which includes difference as a variant, but making it the thought and the production of the 'absolutely different'; making it so that repetition is, for itself, difference in itself" (94). This futural repetition, however, must be distinguished from another kind that similarly treats the past as a default condition and the present agent as a being to be overcome. Deleuze cites Kierkegaard in particular, but also Pascal and Péguy, as thinkers who "were not ready to pay the necessary price" (95) insofar as they counseled a choice to overcome oneself that ultimately "is supposed to restore everything to us" (Deleuze 1986, 116). To the list can be added Heidegger, insofar as he links repetition to authenticity: "when historicality is authentic, it understands history as the 'recurrence' of the possible, and knows that a possibility will recur only if existence is open for it fatefully, in a moment of vision, in resolute repetition" (Heidegger 1962, 444). Indeed, Deleuze contends, Heidegger's critique of Nietzsche's eternal return suggests that he does not fully disengage difference from its subordination to identity and representation (Deleuze 1994, 66; on this point see also Kuiken 2005).

If the imperative to become what one is has any meaning, it cannot take the form of finding an authentic self. In rejecting such authenticity, the thought of eternal return presents the self as an out-of-sync and dissymmetrical multiplicity that generates illusions of stability and centeredness. This is not a form of nihilism, since this immanent synthesis is not lacking in sense, but rather expresses the sense of becoming as such. Nor is it an attempt to escape time and the past, even if it is a call for overcoming and transmutation. When Nietzsche counsels "giving style" to one's character, he does not propose an aesthetic creation *ex nihilo* but rather a strategic deployment and, where possible, a reworking of existing material, all in accordance with a single taste.[13] This material includes those seemingly enduring but ultimately superficial identities and oppositions that purport to give meaning and sense to one's self and one's world but are insufficient to the self's and world's difference and discontinuity. But this reworking is possible only with a revaluation of values, a move beyond the morality of good and evil, and redemption from guilty conscience. "For one thing is needful," Nietzsche writes, "that a human being should *attain* satisfaction with himself, whether it be by means of this or that poetry and art. . . . Whoever is dissatisfied with himself is continually ready for revenge, and we others will be his victims, if only by having to endure his ugly sight" (1974, §290). This task of attaining satisfaction with oneself is clearly ethical, but it is also political—or, rather, micropolitical—insofar as it demands engagement and negotiation with the constitutive temporal relations residing "beneath" identity.

13. "To 'give style' to one's character—a great and rare art! It is practiced by those who survey all the strengths and weaknesses of their nature and then fit them into an artistic plan until every one of them appears as art and reason and even weaknesses delight the eye. Here a large mass of second nature has been added; there a piece of original nature has been removed—both times through long practice and daily work at it. Here the ugly that could not be removed is concealed; there it has been reinterpreted and made sublime. Much that is vague and resisted shaping has been saved and exploited for distant views; it is meant to beckon toward the far and immeasurable. In the end, when the work is finished, it becomes evident how the constraint of a single taste governed and formed everything large and small. Whether this taste was good or bad is less important than one might suppose, if only it was a single taste!" (Nietzsche 1974, §290; see also §299 on artistic strategies for the self).

17

Micropolitics "Beneath" Identity

DESPITE THEIR FICTITIOUSNESS, identity and opposition do structure a certain level of political and social life, figuring most prominently in the standards of normality and deviancy that seem to give sense to various practices and institutions. Deviation from the norm is considered a failure to achieve standards and a falling away from the norm into its opposite. Different standards operate in different domains, yet in all cases social forces are exercised in the name of policing and correcting deviance and compelling conformity with the norms, often using both carrot and stick approaches. Given the hierarchical and exclusionary nature of these oppositional categories—being declared insane or delinquent, for example, certainly excludes individuals from various freedoms and opportunities—a politics of resistance in the name of the marginalized might seem appropriate. Nevertheless, despite the efficacy of identity in these domains, this level is the most superficial one. Its standards are false markers, whose stability and seemingly clear boundaries are merely simulated, and individuals never really embody the model of normality (Deleuze 1995, 173). Moreover, the oppositional structure itself remains abstract and reductive, unable to grasp the complex dynamic of immanent constitutive relations underpinning it. It is therefore necessary, Deleuze and Guattari argue, to distinguish majorities, minorities, and becoming-minor or minoritarian (Deleuze and Guattari 1987, 106–7). With respect to the last, it is not a matter of reversing the relationship between majority and minority or between norm and marginal but of instigating a politics that surmounts these crude divisions through a creative and revolutionary becoming (Deleuze 1995, 170–71). Because oppositional categories are not pregiven, they refer to relations of strife and disjunction from which they emerge. Our social and political worlds thus consist of many layers that cannot be fully separated from one another but must not be confused—levels that create possibilities and necessities for understanding power, knowledge, change, and politics beyond the strictures of identity. This fact opens up a domain of micropolitics, which Deleuze and Guattari's political thought is concerned to trace.

There is, first, a molar level of segmentarity, comprising diverse domains such as family, school, and workplace and classifications such as childhood, adulthood, and homosexuality. At this level of sharp divisions, the various facets of one's complex identity are delineated. One assumes different roles within and across segments, but only one at a time or within specific relations: one passes from childhood to adulthood and from school to work; one can be a father, son, and husband simultaneously, but to different people; one works from nine to five and is a pensioner after age sixty-five. Segments are thereby organized as binaries, even where more than two options are available.

> Segments depend on binary machines which can be very varied if need be. Binary machines of social classes; of sexes, man-woman; of ages, child-adult; of races, black-white; of sectors, public-private; of subjectivations, ours-not ours. These binary machines are all the more complex for cutting across each other, or colliding against each other, confronting each other, and they cut us up in all sorts of directions. And they are not roughly dualistic, they are rather dichotomic: they can operate diachronically (if you are neither *a* nor *b*, then you are *c*: dualism has shifted, and no longer relates to simultaneous elements to choose between, but successive choices; if you are neither black nor white, you are a half-breed; if you are neither man nor woman, you are a transvestite: each time the machine with binary elements will produce binary choices between elements which are not present at the first cutting-up. (Deleuze and Parnet 1987, 128)

Regardless of the choices or assignments made, the segments, despite their diversity, are ominously the same—undoubtedly because they all function on a disciplinary model that polices standards of normal identity (see Deleuze 1995, 169–82).

Segments, however, rest on another micro- or molecular level, where power relations constitute the standards that "code" each particular domain. But this same power, while constituting "normal" identities, also produces opposing forms of madness, delinquency, and perversion. On this dialectical level of constitutive power,[1] resistances take the form of marginals who oppose the coding that depreciates them—yet something of these marginals exists in everyone. They persist as a pervasive and potentially revolutionary element within the segments, always threatening to overturn them.

But opposition refers to a more concrete form of difference and so to another molecular level of strife and flux, one that is affirmative in the Nietzschean sense. This is a level not of power but of what Deleuze and Guattari call desiring-machines. The desire of these machines is not a negative relation to an object in either a Freudian or a Lacanian sense but resembles more the freely mobile processes that Freud locates beyond the pleasure principle but quickly absorbs into a binary schema of instincts. Machinic desire includes a body without organs, or "BwO," a differentiator that "is produced . . . in the connective synthesis" (Deleuze and Guattari 1983, 8) and that repels the components it brings together because they remain unbearable to it (9), making friction and strife always part of these machines. A desiring-machine is an assemblage (*agencement*) that functions off of its own friction. In contrast to the segments and binaries of other levels, the level of desiring-machines consists of chaotic and excessive "lines of flight," which "deterritorialize" molar formations and their sharp divisions. The second level of power and resistance can deterritorialize only partially, since its resistances work in simple opposition to power (Deleuze and Parnet 1987, 136). But complete deterritorializations are possible at the third level because lines of flight break down molar formations not by opposing them but simply by exceeding them, operating in an asystematic rather than an antisystematic way (see Kazarian 1998). This final level is in some sense outside the others, insofar as they cannot fully capture it; but the three levels really "are immanent, caught up in one another" (Deleuze and Parnet 1987, 125), and form a single "assemblage"—a multiplicity of heterogeneous but inseparable realms flowing into and through one another (Deleuze and Guattari 1987, 4). Microlevels of desire and power constitute the molar segments that seem to fix and incorporate them, "reterritorializing" desiring lines of flight, while lines of flight continue within molar organizations, exceeding and dissolving them.

Molecular and molar forms of power code differences within the various

---

1. Deleuze and Guattari (1987, 530–31n39) questionably attribute this dialectical sort of power relation to Foucault, criticizing him for failing to locate a disjunctive level of desire that is more primary than any opposition of power and resistance (for a more sustained and subtle account of this criticism, see Deleuze 1997a). This reading and critique, however, are absent from Deleuze's (1988) monograph on Foucault.

segments, giving each domain markers of normality and deviance. Domains are independent, yet they also communicate and pass their subjected products to one another: school disciplines children in preparation for work; courts and prisons use different norms and practices yet also resonate with each other, the first producing convicts through judgments of guilt passed on acts and the second receiving convicts and turning them into delinquents who usually reoffend and end up back in court. Every domain similarly refers beyond itself, while the State enables these diverse domains to communicate. The State "overcodes" the coded domains by imposing its own rigid segmentarity on the segments (Deleuze and Guattari 1987, 209–10). Or, rather, the State is the realization of this overcoding (223; see also Deleuze and Parnet 1987, 129). It regulates transfers rather than reduces segments to one homogeneous blob, making the State an important but limited site of political struggle: the problem of crime, for example, cannot be adequately dealt with at the State level alone, because it is a relay rather than an all-or-nothing power holder. Moreover, the State and its institutions of coding are always pitted against deterritorializing excesses. They maintain an uneasy alliance with the forces of capitalism, which do not code differences but rather submit them to an axiomatic of exchange value. Capitalism effects a partial deterritorialization of the State, but it is also implicated in forms of reterritorialization. More profoundly, the State is challenged by a chaotic excess of desire that Deleuze and Guattari call the "war machine." It lies outside the State apparatus and is only partially integrated through military and police institutions, which give this constitutive aggressiveness a defined purpose: "*The State has no war machine of its own;* it can only appropriate one in the form of a military institution, one that will continually cause it problems" (Deleuze and Guattari 1987, 355). Chaos must not be conceived from the State's perspective, which considers it the polar opposite of order, a Hobbesian state of nature. Chaos is a realm of events, a nonsense that constitutes sense, and so consists not of random accidents but rather of "nomadic" passages through a second-order difference. These movements of chaos are as much collective as they are personal (Deleuze and Parnet 1987, 127). While they may appear stationary from the perspective of identity and opposition, these lines of flight, moving with "infinite speed," can initiate "a curious stationary journey" (127).

This multilayered complex of movements and codings calls for a multilayered politics and recognition of the way all things, at whatever level, are political. The personal is political, not simply because the barrier between public and private is always drawn in the public or political realm, nor because public and private are both organized by large-scale powers such as capitalism, patriarchy, racism, and heterosexism. It is rather because any thing, in the molecular fluxes that constitute it and in which it participates, and in the creative deterritorializations it can enact, effects a reification or an overcoming of formations of identity and opposition. And it is because these social and political formations, being constituted by historically specific and contingent practices, are nonetheless also tied to an excess and discontinuity that give them their out-of-sync structure.

Politics must therefore involve more than reform of existing institutions or even revolutionary opposition to them. Every level of politics must include creative and experimental strategies. A politics directed toward the segments may seek to modify or reform them, perhaps even radically. Yet these are not straightforward procedures, as they "testify to a long labour which is not merely aimed against the State and the powers that be, but directly at ourselves" (Deleuze and Parnet 1987, 138). On the one hand, it is not a matter of simply making rigid segments more flexible, "believing that a little suppleness is enough to make things 'better'" (Deleuze and Guattari 1987, 215). On the other hand, segmentation is undeniably necessary: "the segments which run through us and through which we pass are . . . marked by a rigidity which reassures us, while turning us into creatures which are the most fearful, but also the most pitiless and bitter. . . . Even if we had the power to blow it up, could we succeed in doing so without destroying ourselves, since it is so much a part of the conditions of life, including our organism and our very reason?" (Deleuze and Parnet 1987, 138). The realm of constitutive relations suggests a politics of the marginal and the revolutionary that seeks to overturn ingrained, exclusionary standards. But here lies a danger of "microfascism," which reinforces blunt oppositions through a spiteful friend/enemy politics common to both segmented societies and the resistance movements that challenge them. Microfascisms are rife in modern societies, both liberal and totalitarian, and stretch across their political spectrums. They are part of a human, all too human thinking that fails to affirm difference and acknowledge the simulated status of identity.

The realm of desiring-machines therefore calls for a third kind of politics, which concerns neither reform nor opposition but literally "doing something different." There is a kind of experimentation, Deleuze and Guattari argue, that standard political theory might not consider political but that is eminently political in its power to surmount the categories of standard politics. At this level politics is a question of how individuals and collectives can overcome the identities and oppositions that seem to exhaust their meaning and sense by instituting deterritorializing lines of flight. It is a matter, De-leuze and Guattari say, of making oneself a body without organs or BwO (1987, 149–66) by disaggregating the various elements and relations that organize oneself into the segmented and stratified identity one assumes. As the concrete sense of an individual or collective necessarily refers to heterogeneous axes of difference, the self must be understood as a divergent assemblage of relations, which include the material, linguistic, human, ani-mal, and visceral: "For the BwO is all of that: necessarily a Place, necessarily a Plane, necessarily a Collectivity (assembling elements, things, plants, ani-mals, tools, people, powers, and fragments of all of these; for it is not 'my' body without organs, instead the 'me' (*moi*) is on it, or what remains of me, unalterable and changing in form, crossing thresholds)" (161). The BwO is an experiment in the opportunities for mutation that this complex but seemingly sedimentary structure provides.

> This is how it should be done: Lodge yourself on a stratum, experi-ment with the opportunities it offers, find an advantageous place on it, find potential movements of deterritorialization, possible lines of flight, experience them, produce flow conjunctions here and there, try out continuums of intensities segment by segment, have a small plot of new land at all times. It is through a meticulous relation with the strata that one succeeds in freeing lines of flight, causing conjugated flows to pass and escape and bringing forth continuous intensities for a BwO. (161)

To strive in this way beyond the crude divisions established between the human and the nonhuman, the male and the female, the normal and the deviant, and so forth, toward more subtle and complex relations of differ-ence is to engage in a politics that seeks to overcome categories that have been treated as necessities but whose stability and substantiality are in fact no more than optical illusions.

The dangers of this level come largely from the constitutive nature of desiring-machines, which can either affirm disjunction or reinstate opposi-tion. Lines of flight may connect productively or may collapse into isolated and empty "black holes"; they may fall into a trap of clarity where fascism reemerges as a dogmatic certainty of "the truth"; or a line of flight may become a line of self-destruction, and if this form of desire takes control of the State, it may become macroscopic fascism: "in fascism, the State is far less totalitarian than it is *suicidal*. There is in fascism a realized nihilism"

(Deleuze and Guattari 1987, 230; see 227–31 generally). Similarly, experimental BwOs can be botched (149). They may become empty, cancerous, or fascist (163). The question is "knowing whether we have it within our means to make the selection, to distinguish the BwO from its doubles: empty vitreous bodies, cancerous bodies, totalitarian and fascist" (165). Deleuze and Guattari ask but arguably never answer this question. Deleuze, for example, merely says: "There is no general prescription. We have done [*sic*] with all globalizing concepts" (Deleuze and Parnet 1987, 144). But perhaps this is the only possible answer, since uncertainty is what makes the body without organs both experimental and political. The BwO is a matter of both political thought and political practice, which are themselves brought together in relations of mutual imbrication and difference. Absent foundational standards, the construction of the BwO is necessarily a matter of pragmatism and strategy. Through our thought and practice at this level, Deleuze and Guattari argue, we seek to negotiate the impasses imposed on us by the very identities and oppositions that seem to give us structure and sense but are ultimately inadequate to our lives.

18

The Care of the Self and Politics

THE WILL TO TRUTH underpinning disciplinary and normalizing powers may seek to categorize differences according to standards of normality and deviance. But the operations of these power relations always produce something else. Moreover, the frictions and conflicts within disciplinary institutions and practices mean that they operate, from the will to truth's perspective, only by breaking down.[1] Nevertheless, a regularity persists, giving sense and direction to the entire disciplinary regime. This regularity is found in the schema of correspondence connecting the heterogeneous domains of desire and truth, which constitutes desire as a hidden source of truth about the self. It is against the backdrop of this regularity that Foucault closes the first volume of *The History of Sexuality* by suggesting the possibility of "a different economy of bodies and pleasures" (1990a, 159). It is a mistake to see this as a move to place bodies and pleasures outside of power and discourse.[2] It is rather part of a strategy thoroughly embedded in discourse, targeting a link central to modern discipline. Bringing to the fore the idea that what has been forgotten by modern disciplinary policing is the pleasure of the sexual act,[3] it seeks not an escape from this game of truth but instead a way of "playing it otherwise" (Foucault 1988, 15).

Does this leave nothing more than different, historically contingent configurations and senses of power and truth that cannot be judged better or

---

1. This may be compared with the functioning of Deleuze and Guattari's desiring-machines, which "work only when they break down, and by continually breaking down" (Deleuze and Guattari 1983, 8).

2. Examples of this charge against Foucault include Butler (1990, 93–111) and Newman (2000, chapter 4). A detailed and sophisticated response is provided by McWhorter (1999).

3. "And I could say that the modern 'formula' is desire, which is theoretically underlined and practically accepted, since you have to liberate your own desire. Acts are not very important, and pleasure—nobody knows what it is!" (Foucault 1984a, 359).

worse with respect to one another? That conclusion underplays the import of archaeological and genealogical analyses. Meaning and knowledge may be impossible without disjunctive linkages, making it impossible to escape power relations and perhaps also the simulacral unities and identities arising from them. Nevertheless, by uncovering the microscopic dispersion beneath these unities, another sense can be glimpsed that is overlooked by representational thought and the will to truth. Or, perhaps better, it is possible to glimpse the sense of sense, or the sense of sensible statements as such, which underlies all statements and allows them to "make sense." This second order of sense is the paradoxical but nonetheless positive content of a dispersion that exceeds the terms of identity and opposition and that is ironed out by oppositional thought. Rather than conceiving Otherness in the terms prescribed by the will to truth—as either an oppositional difference compatible with identity or an excessive, unmediatable difference that may be elevated to divinity or reduced to chaos and materiality, but in either case is treated as a difference lying outside the boundaries of identity—this sense presents it as the immanent differenciator that disjoins differences, folding together heterogeneous but mutually imbricated domains.[4] This untimely excess arises with discursive practices, problematizing the surface identities and oppositions that coordinate these practices. In functioning this way, however, this excess can also help us modify our practices and ourselves.

Foucault's final turn to practices of the self cannot be understood without taking this excess into account, in terms of both the way it exceeds us by traversing and cutting us apart, demolishing anything within us that might be seen as a homogeneous substance, and the way it leads us to exceed ourselves. The self-to-self relation, "by which the individual constitutes and recognizes himself *qua* subject" (Foucault 1992, 6), is hardly a realm of freedom divorced from power. The strategies and practices available for the self to train itself are products of games of truth and the meanings and codes they establish. The self-to-self relation, in turn, resides within a microscopic realm where selves are constituted through power relations that interpenetrate them and within which they participate. Nonetheless, because these power relations are relations of disjunction, dispersion, and strife, this micropolitical realm is as much a realm of self-creation, self-stylization, and self-experimentation as it is a realm of self-discipline and training.

Because moral codes presuppose a self that fashions and positions itself in relation to these codes, this self cannot have the form of a moral sub-

---

4. This excess is worked out in relation to Foucault's ethical works in Golding (1995).

ject—a self-reflexive "I" or ego that separates itself from what it is not and takes responsibility for itself—that is supposed to be the end product of this transformation. It must instead take the form of a unity of dispersion, a complex of heterogeneities with no firm distinction between inside and outside. Even if the ancient world was no golden age (see Foucault 1984a, 344–51), its ethics provides a picture of what this self that is more nebulous than a moral subject might be. Self-mastery implies a difference within the self, but absent a Christian judgment that underpins practices of self-sacrifice,[5] this difference does not take the form of a hierarchical opposition between a naturally pure self and a sinful and corrupt Other. Rather, as with the Greeks' agonistic conception of the self, "the adversary that was to be fought, however far removed it might be by nature from any conception of the soul, reason, or virtue, did not represent a different, ontologically alien power" (Foucault 1992, 68). This "internal" complexity of the self is matched by a multifaceted relation to its environment, which is seen, for example, in the way medical appraisals of the sexual act developed subtle measurements of the diverse factors constituting the act, including social and environmental factors and character traits (see Foucault 1990b, 116–17). The self portrayed in this ethics is therefore a nexus where socially constituted rules, practices, techniques, and institutions, along with bodies, pleasures, desires, and memories, all converge, none of these being reducible to the others nor strictly separable from them.[6]

This lack of homogeneity in the self and Foucault's step back from morality to ethics would be redundant, however, if the practices of the self did no more than fix it within an identity or train it to become a subject conforming to a moral code. Any moral system, however, being constituted by power relations, is replete with points of friction and problematization. For the Greeks, the status of boys is ethically problematic, for even though love of one's own sex and love of the other are not seen as opposites, the sex act itself is considered in terms of an opposition between activity and passivity, neither of which is suitable for the boy. For the Romans, the issue of pederastic love becomes less important, but the care of the self becomes problematic because ethical training is no longer tied to the status of the freeman who must prepare for rule in the polis, leading to the prominent alternatives

5. This is the kind of self-sacrifice underpinning Christian practices of *exomologesis* and *exogoreusis* (Foucault 1999).

6. This conception of the self as a nexus of interpenetrating material, social, and visceral relations has been prominently developed in political theory by William E. Connolly (see, in particular, Connolly 2002).

of renewed political activism or a withdrawal into political apathy (see Foucault 1990b, 37–68, 189–232; 1992, 185–225). In both cases, ethical practice becomes more experimental and self-stylization becomes more subtle. Ethics becomes the negotiation of points of moral ambiguity, where the oppositions that structure moral systems reveal their inadequacy and available alternatives cannot be easily separated into good and bad or right and wrong. But it also becomes a process of creative self-overcoming on an ethical level "underneath" the largely stable level of moral codes.[7] The Greeks used the problem of the love of boys to expand themselves beyond what they were, creating new ways to relate to another, to understand the freedom of the other, and thereby to understand themselves.[8] Similarly, the problematization of the self that arose with the decline of the polis compelled shifts in the ethical relations to one's spouse and one's society, as well as to oneself.

In these ways Foucault's later works develop the ethical themes already found in his earlier calls to combat fascism within the self (see Foucault's introduction to Deleuze and Guattari 1983) and to think beyond the idea of one's adversary as an enemy (Foucault 1991, 180–81). In those cases too it is a problem of surpassing the binary logics that define the contours of identity and developing an ethics of overcoming by moving beyond conceptions of the homogeneous self. The dimension of ethical negotiation is always present owing to the disjunctive nature of both power and the self and because of the excess that always accompanies political and ethical practices. Yet it is perhaps not surprising that the care of the self and the micropolitics it entails are not often thematized by many who engage with Foucault, since these ideas do not fit well into more conventional understandings of power and politics.[9] Such understandings regularly link identity and politics, often maintaining that the establishment of a collective subject through the mobilization of identity—whether this identity is considered pregiven and necessary or historically contingent—is a precondition for political action.[10] While

7. "I am not supposing that the codes are unimportant. But one notices that they ultimately revolve around a rather small number of rather simple principles: perhaps men are not much more inventive when it comes to interdictions than they are when it comes to pleasures. Their stability is also rather remarkable; the notable proliferation of codifications . . . occurred rather late in Christianity. On the other hand, it appears . . . that there is a whole rich and complex field of historicity in the way the individual is summoned to recognize himself as an ethical subject of sexual conduct" (Foucault 1992, 32).

8. I owe this point to an excellent paper by Matthew Hammond (2003).

9. Exceptions to this neglect include Connolly (2002), Deleuze (1988, 94–123), and Golding (1995).

10. In many contemporary political theories that aspire to move beyond naturalized or essentialized identities, explicit caveats continue to stress the political and even ontological

these views often accept that a plurality of identities must be incorporated into this collective and that the ultimate fixing of identities is impossible, the need for such a marker and the concomitant need to negotiate its inevitable failure still set the horizon of their political imaginations. As is well known, the erection of these boundaries that separate such collectives from what they are not comes at the cost of erasing numerous complexities on both sides of the divides, in many ways repeating the violence of the forms of power being contested. But these uses of identity also fundamentally miss the way power structures meaning and sense, because they accept as substantial the illusions generated by a dynamic of dispersion. Often these erasures and abstractions are still considered necessary, however, on the grounds that any positive content without such negative relations would be fanciful and would effect a flawed return to the idea of something outside of power, discourse, and relationality.

When Foucault accepts the idea that his work is an ethics that feeds into politics (Foucault 1984c, 375), he suggests that this is not the way the game of politics needs to be played. If we leave behind the idea that politics requires in the first instance the construction of an efficacious political identity—a move compelled by considerations of the concrete structure of time—what emerges is a micropolitical domain of ethical negotiation where what matters is not the ability to construct an identity but rather the capacity for revaluations that move us beyond crude oppositions. The sort of positivity often criticized by those who link identity and politics is one firmly entrenched in the logic of identity and the will to truth: it is the positive difference that Hegel shows to be an abstraction and dissolves through negation and sublation. But Foucault articulates a different kind of positivity of power, discourse, and the self, which sustains relationality but moves it from the negativity of a still abstract opposition toward an immanent multiplicity. When we "withdraw allegiance from the old categories of the Negative" by turning groups and collectives into "a constant generator of de-individualization" (Foucault's introduction to Deleuze and Guattari 1983, xiii–xiv), we move beyond the need to ground politics in simulacral identities and subjectivities. Although Foucault's work was always concerned with these categories of identity, his lesson about them, ironically, is that they are not as important as we usually think.

---

need for some moment of identity or identification. Examples include Butler (1993, 117–18), Laclau (1990, 90), and Laclau and Mouffe (1985, 112). The differences between recent Lacan-inspired and Deleuze-inspired political theories, including their different orientations to identity, are explored by Tønder and Thomassen (2005).

REFERENCES

Adalier, Gokhan. 2001. "The Case of *Theaetetus*." *Phronesis* 46 (1): 1–37.

Adorno, Theodor. 1978. *Minima Moralia: Reflections from Damaged Life*. Trans. E. F. N. Jephcott. London: Verso.

———. 1995. *Negative Dialectics*. Trans. E. B. Ashton. New York: Continuum.

Adorno, Theodor, and Max Horkheimer. 1997. *Dialectic of Enlightenment*. Trans. John Cumming. London: Verso.

Agamben, Giorgio. 2002. "Absolute Immanence." In *An Introduction to the Philosophy of Gilles Deleuze*, ed. Jean Khalfa, 151–69. London: Continuum.

Aitken, Andrew. 2004. "An 'Applied Rationalism' of Time: A Re-investigation of the Relationship Between Bachelard and Bergson-Deleuze." *Pli: The Warwick Journal of Philosophy* 15: 76–102.

Al-Saji, Alia. 2004. "The Memory of Another Past: Bergson, Deleuze, and a New Theory of Time." *Continental Philosophy Review* 37 (2): 203–39.

Allen, R. E., ed. 1965. *Studies in Plato's Metaphysics*. New York: Humanities Press, and London: Routledge & Kegan Paul.

Althusser, Louis. 1984. "Ideology and the Ideological State Apparatuses (Notes Towards an Investigation)." In Althusser, *Essays on Ideology*, 1–60. London: Verso.

———. 1996. *For Marx*. Trans. Ben Brewster. London: Verso.

Ansell Pearson, Keith. 1999. *Germinal Life: The Difference and Repetition of Deleuze*. London: Routledge.

———. 2002. *Philosophy and the Adventure of the Virtual: Bergson and the Time of Life*. London: Routledge.

Aristotle. 1933–35. *Metaphysics*. Trans. Hugh Tredennick. 2 vols. Cambridge, Mass.: Loeb Classics.

———. 1934–57. *Physics*. Trans. P. H. Wicksteed and F. M. Cornford. 2 vols. Cambridge, Mass.: Loeb Classics.

———. 1984a. *Categories*. In *The Complete Works of Aristotle: The Revised Oxford Translation*, ed. Jonathan Barnes, 2 vols., 1:3–26. Princeton: Princeton University Press.

———. 1984b. *On the Soul*. In *The Complete Works of Aristotle: The Revised Oxford Translation*, ed. Jonathan Barnes, 2 vols., 1:641–92. Princeton: Princeton University Press.

Augustine. 1961. *Confessions*. Trans. R. S. Pine-Coffin. London: Penguin Books.

Bachelard, Gaston. 2000. *The Dialectic of Duration*. Trans. Mary McAllester Jones. Introduction by Cristina Chimisso. Manchester, UK: Clinamen Press.

Badiou, Alain. 2000. *Deleuze: The Clamor of Being*. Trans. Louise Burchill. Minneapolis: University of Minnesota Press.

Baudrillard, Jean. 1993. *Symbolic Exchange and Death*. Trans. Iain Hamilton Grant. Thousand Oaks, Calif.: Sage.

Bennett, Jane. 2001. *The Enchantment of Modern Life: Attachments, Crossings, and Ethics*. Princeton: Princeton University Press.

Bergson, Henri. 1910. *Time and Free Will: Essays on the Immediate Data of Consciousness*. Trans. F. L. Pogson. London: George Allen and Unwin.

———. 1956. *The Two Sources of Morality and Religion*. Trans. R. Ashley Audra and Cloudesley Brereton, with W. Horsfall Carter. Garden City, N.J.: Doubleday.

———. 1970. "Aristotle's Concept of Place." In *Studies in Philosophy and the History of Philosophy*, vol. 5, *Ancients and Moderns*, ed. John K. Ryan, 20–71. Washington, D.C.: Catholic University of America Press.

———. 1983. *An Introduction to Metaphysics: The Creative Mind*. Trans. Mabelle L. Andison. Totowa, N.J.: Rowman & Allanheld.

———. 1991. *Matter and Memory*. Trans. N. M. Paul and W. S. Palmer. New York: Zone Books.

———. 1998. *Creative Evolution*. Trans. Arthur Mitchell. Mineola, N.Y.: Dover Publications.

———. 1999. *Duration and Simultaneity: Bergson and the Einsteinian Universe*. Ed. Robin Durie, trans. Leon Jacobson, with Mark Lewis and Robin Durie. Manchester, UK: Clinamen Press.

Boeri, Marcelo D. 2001. "The Stoics on Bodies and Incorporeals." *Review of Metaphysics* 54 (4): 723–52.

Boothby, Richard. 1991. *Death and Desire: Psychoanalytic Theory in Lacan's Return to Freud*. New York: Routledge.

———. 2001. *Freud as Philosopher: Metapsychology After Lacan*. New York: Routledge.

Borradori, Giovani. 2001. "The Temporalization of Difference: Reflections on Deleuze's Interpretation of Bergson." *Continental Philosophy Review* 34 (1): 1–20.

Bostock, David. 1991. "Aristotle on Continuity in *Physics* VI." In *Aristotle's Physics: A Collection of Essays*, ed. Lindsay Judson, 179–212. Oxford: Clarendon Press.

Boundas, Constantin V. 1996. "Deleuze-Bergson: An Ontology of the Virtual." In *Deleuze: A Critical Reader*, ed. Paul Patton, 81–106. Oxford: Blackwell.

Boyer, Carl B. 1991. *A History of Mathematics*. 2d ed. Rev. Uta C. Merzbach. New York: John Wiley & Sons.

Braidotti, Rosi. 1994. "Toward a New Nomadism: Feminist Deleuzean Tracks; or, Metaphysics and Metabolism." In *Gilles Deleuze and the Theater of Philosophy*, ed. Constantin V. Boundas and Dorothea Olkowski, 157–86. London: Routledge.

Bréhier, Émile. 1997. *La théorie des incorporels dans l'ancien stoïcisme*. 9th ed. Paris: Librairie Philosophique J. Vrin.

Butler, Judith. 1987. *Subjects of Desire: Hegelian Reflections in Twentieth-Century France*. New York: Columbia University Press.

———. 1990. *Gender Trouble: Feminism and the Subversion of Identity*. New York: Routledge.

———. 1993. *Bodies That Matter: On the Discursive Limits of "Sex."* New York: Routledge.

Cantor, Georg. 1955. *Contributions to the Founding of the Theory of Transfinite Numbers*. Trans. Philip E. B. Jourdain. New York: Dover Publications.

Connolly, William E. 1993. *Political Theory and Modernity*. 2d ed. Ithaca: Cornell University Press.

———. 2002. *Neuropolitics: Thinking, Culture, Speed*. Minneapolis: University of Minnesota Press.

Coope, Ursula. 2001. "Why Does Aristotle Say That There Is No Time Without Change?" *Proceedings of the Aristotelian Society* 101 (3): 359–67.

———. 2005. *Time for Aristotle: Physics IV.10–14*. Clarendon: Oxford University Press.

Cornford, Francis M. 1935. *Plato's Theory of Knowledge*. London: Routledge & Kegan Paul.

Dedekind, Richard. 1963. *Essays on the Theory of Numbers*. Trans. Wooster Woodruff Beman. New York: Dover Publications.

Deleuze, Gilles. 1983. *Nietzsche and Philosophy*. Trans. Hugh Tomlinson. London: Athlone Press.

———. 1984. *Kant's Critical Philosophy: The Doctrine of the Faculties*. Trans. Hugh Tomlinson and Barbara Habberjam. London: Athlone Press.

———. 1986. *Cinema 1: The Movement-Image*. Trans. Hugh Tomlinson and Barbara Habberjam. London: Athlone Press.

———. 1988. *Foucault*. Trans. Seán Hand. London: Athlone Press.

———. 1989. *Cinema 2: The Time-Image*. Trans. Hugh Tomlinson and Robert Galeta. London: Athlone Press.

———. 1990. *The Logic of Sense*. Trans. Mark Lester, with Charles Stivale. New York: Columbia University Press.

———. 1991. *Bergsonism*. Trans. Hugh Tomlinson and Barbara Habberjam. New York: Zone Books.

———. 1993. *The Fold: Leibniz and the Baroque*. Trans. Tom Conley. Minneapolis: University of Minnesota Press.

———. 1994. *Difference and Repetition*. Trans. Paul Patton. London: Athlone Press.

———. 1995. *Negotiations*. Trans. Martin Joughin. New York: Columbia University Press.

———. 1997a. "Desire and Pleasure." In *Foucault and His Interlocutors*, ed. A. I. Davidson, 183–92. Chicago: University of Chicago Press.

———. 1997b. "Review of Jean Hyppolite." In Jean Hyppolite, *Logic and Existence*, trans. Leonard Lawlor and Amit Sen, 191–95. Albany: State University of New York Press.

———. 1998. *Essays Critical and Clinical*. Trans. Daniel W. Smith, with Michael A. Greco. London: Verso.

———. 1999. "Bergson's Conception of Difference." Trans. Melissa McMahon. In *The New Bergsonism*, ed. John Mullarkey, 42–65. Manchester: Manchester University Press.

———. 2006. *Two Regimes of Madness: Texts and Interviews, 1975–1995*. Ed. David Lapoujade. Trans. Ames Hodges and Mike Taormina. New York: Semiotext(e).

Deleuze, Gilles, and Félix Guattari. 1983. *Anti-Oedipus: Capitalism and Schizophrenia*. Trans. Robert Hurley, Mark Seem, and Helen R. Lane. Minneapolis: University of Minnesota Press.

———. 1987. *A Thousand Plateaus: Capitalism and Schizophrenia*. Trans. Brian Massumi. Minneapolis: University of Minnesota Press.

———. 1994. *What Is Philosophy?* Trans. Hugh Tomlinson and Graham Burchell. New York: Columbia University Press.

Deleuze, Gilles, and Claire Parnet. 1987. *Dialogues.* Trans. Hugh Tomlinson and Barbara Habberjam. New York: Columbia University Press.

Derrida, Jacques. 1978. "From Restricted to General Economy: A Hegelianism Without Reserve." In Derrida, *Writing and Difference,* trans. Alan Bass, 251–77. London: Routledge.

———. 1981. "Plato's Pharmacy." In Derrida, *Dissemination,* trans. Barbara Johnson, 61–171. Chicago: University of Chicago Press.

———. 1982. "*Ousia* and *Grammē:* Note on a Note from *Being and Time.*" In Derrida, *Margins of Philosophy,* trans. Alan Bass, 29–67. Chicago: University of Chicago Press.

Descombes, Vincent. 1980. *Modern French Philosophy.* Trans. L. Scott-Fox and J. M. Harding. Cambridge: Cambridge University Press.

Deutscher, Penelope. 1994. "'The Only Diabolical Thing About Women . . .': Luce Irigaray on Divinity." *Hypatia* 9 (4): 88–111.

Dreyfus, Hubert, and Paul Rabinow. 1982. *Michel Foucault: Beyond Structuralism and Hermeneutics.* Hemel Hempstead, UK: Harvester Wheatsheaf.

Durie, Robin. 2000. "Splitting Time: Bergson's Philosophical Legacy." *Philosophy Today* 40 (summer): 152–68.

———. 2004. "The Mathematical Basis of Bergson's Philosophy." *Journal of the British Society for Phenomenology* 35 (1): 54–67.

Ellenberger, Henri F. 1970. *The Discovery of the Unconscious: The History and Evolution of Dynamic Psychiatry.* New York: Basic Books.

Faulkner, Keith W. 2006. *Deleuze and the Three Syntheses of Time.* New York: Peter Lang.

Foucault, Michel, ed. 1975. *I, Pierre Rivière, Having Slaughtered My Mother, My Sister, and My Brother: A Case of Parricide in the 19th Century.* Trans. Frank Jellinek. Lincoln: University of Nebraska Press.

———. 1977a. "Intellectuals and Power." In Foucault, *Language, Counter-Memory, Practice: Selected Essays and Interviews,* trans. Donald F. Bouchard and Sherry Simon, ed. Donald F. Bouchard, 205–17. Ithaca: Cornell University Press.

———. 1977b. "Revolutionary Action: 'Until Now.'" In Foucault, *Language, Counter-Memory, Practice: Selected Essays and Interviews,* trans. Donald F. Bouchard and Sherry Simon, ed. Donald F. Bouchard, 218–33. Ithaca: Cornell University Press.

———. 1977c. "Theatricum Philosophicum." In Foucault, *Language, Counter-Memory, Practice: Selected Essays and Interviews,* trans. Donald F. Bouchard and Sherry Simon, ed. Donald F. Bouchard, 165–96. Ithaca: Cornell University Press.

———. 1979. *Discipline and Punish: The Birth of the Prison.* Trans. Alan Sheridan. New York: Vintage Books.

———. 1980a. *Herculine Barbin: Being the Recently Discovered Memoirs of a Nineteenth-Century French Hermaphrodite.* Trans. Richard McDougall. New York: Pantheon Books.

———. 1980b. *Power/Knowledge: Selected Interviews and Other Writings, 1972–1977.* Trans. Colin Gordon, Leo Marshall, John Mepham, and Kate Soper, ed. Colin Gordon. Brighton: Harvester.

———. 1982. "Afterword: The Subject and Power." In Hubert Dreyfus and Paul Rabinow, *Michel Foucault: Beyond Structuralism and Hermeneutics,* 208–26. Hemel Hempstead, UK: Harvester Wheatsheaf.

————. 1984a. "On the Genealogy of Ethics: Overview of a Work in Progress." In *The Foucault Reader*, ed. Paul Rabinow, 340–72. New York: Pantheon Books.

————. 1984b. "The Order of Discourse." In *Language and Politics*, ed. Michael Shapiro, 108–38. New York: New York University Press.

————. 1984c. "Politics and Ethics: An Interview." In *The Foucault Reader*, ed. Paul Rabinow, 373–80. New York: Pantheon Books.

————. 1988. "The Ethic of Care for the Self as a Practice of Freedom." In *The Final Foucault*, ed. James Bernauer and David Rasmussen, 1–20. Cambridge: MIT Press.

————. 1989a. *The Archaeology of Knowledge*. Trans. A. M. Sheridan Smith. London: Routledge.

————. 1989b. *The Birth of the Clinic: An Archaeology of Medical Perception*. Trans. A. M. Sheridan. London: Routledge.

————. 1989c. *Madness and Civilization: A History of Insanity in the Age of Reason*. Trans. Richard Howard. London: Routledge.

————. 1990a. *The History of Sexuality*. Vol. 1, *An Introduction*. Trans. Robert Hurley. New York: Vintage Books.

————. 1990b. *The History of Sexuality*. Vol. 3, *The Care of the Self*. Trans. Robert Hurley. Harmondsworth: Penguin Books.

————. 1991. *Remarks on Marx: Conversations with Duccio Trombadori*. Trans. R. James Goldstein and James Cascaito. New York: Semiotext(e).

————. 1992. *The History of Sexuality*. Vol. 2, *The Use of Pleasure*. Trans. Robert Hurley. Harmondsworth: Penguin Books.

————. 1999. "About the Beginning of the Hermeneutics of the Self." In *Religion and Culture by Michel Foucault*, ed. Jeremy R. Carrette, 158–81. Manchester: Manchester University Press.

Fraser, Nancy. 1989. *Unruly Practices: Power, Discourse and Gender in Contemporary Social Theory*. Cambridge, UK: Polity Press.

Freud, Sigmund. 1952. *On Dreams*. Trans. James Strachey. London: Hogarth Press and Institute of Psycho-Analysis.

————. 1953. "Three Essays on Sexuality." In *The Standard Edition of the Complete Psychological Works of Sigmund Freud*, ed. and trans. James Strachey, 24 vols., 7:123–245. London: Hogarth Press and Institute for Psycho-Analysis.

————. 1957a. "Beyond the Pleasure Principle." In *The Standard Edition of the Complete Psychological Works of Sigmund Freud*, ed. and trans. James Strachey, 24 vols., 18:3–64. London: Hogarth Press and Institute for Psycho-Analysis.

————. 1957b. "Instincts and Their Vicissitudes." In *The Standard Edition of the Complete Psychological Works of Sigmund Freud*, ed. and trans. James Strachey, 24 vols., 14:109–40. London: Hogarth Press and Institute for Psycho-Analysis.

————. 1957c. "On Narcissism: An Introduction." In *The Standard Edition of the Complete Psychological Works of Sigmund Freud*, ed. and trans. James Strachey, 24 vols., 14:67–102. London: Hogarth Press and Institute for Psycho-Analysis.

————. 1957d. "The Unconscious." In *The Standard Edition of the Complete Psychological Works of Sigmund Freud*, ed. and trans. James Strachey, 24 vols., 14:161–215. London: Hogarth Press and Institute for Psycho-Analysis.

————. 1959. "An Autobiographical Study." In *The Standard Edition of the Complete Psychological Works of Sigmund Freud*, ed. and trans. James Strachey, 24 vols., 20:3–74. London: Hogarth Press and Institute for Psycho-Analysis.

————. 1960. *Totem and Taboo: Some Points of Agreement Between the Mental Lives of Savages and Neurotics*. Trans. James Strachey. London: Ark Paperbacks.

————. 1961a. "The Ego and the Id." In *The Standard Edition of the Complete Psychological Works of Sigmund Freud*, ed. and trans. James Strachey, 24 vols., 19:3–66. London: Hogarth Press and Institute for Psycho-Analysis.

————. 1961b. "Negation." In *The Standard Edition of the Complete Psychological Works of Sigmund Freud*, ed. and trans. James Strachey, 24 vols., 19:233–39. London: Hogarth Press and Institute for Psycho-Analysis.

————. 1961c. "A Note upon the Mystic Writing-Pad." In *The Standard Edition of the Complete Psychological Works of Sigmund Freud*, ed. and trans. James Strachey, 24 vols., 19:225–32. London: Hogarth Press and Institute for Psycho-Analysis.

————. 1961d. "Some Psychical Consequences of the Anatomical Distinction Between the Sexes." In *The Standard Edition of the Complete Psychological Works of Sigmund Freud*, ed. and trans. James Strachey, 24 vols., 19:243–58. London: Hogarth Press and Institute for Psycho-Analysis.

————. 1962. "Screen Memories." In *The Standard Edition of the Complete Psychological Works of Sigmund Freud*, ed. and trans. James Strachey, 24 vols., 3:301–22. London: Hogarth Press and Institute for Psycho-Analysis.

————. 1963. "From the History of an Infantile Neurosis." In *Three Case Histories*, ed. Philip Rieff, 161–280. New York: Collier Books.

————. 1965. *New Introductory Lectures on Psychoanalysis*. Trans. James Strachey. New York: W. W. Norton.

————. 1966. *Introductory Lectures on Psychoanalysis*. Trans. James Strachey. New York: W. W. Norton.

————. 1994. *Civilization and Its Discontents*. Trans. Joan Riviere. New York: Dover Publications.

Freud, Sigmund, and Josef Breuer. 1955. "Studies on Hysteria." *The Standard Edition of the Complete Psychological Works of Sigmund Freud*, ed. and trans. James Strachey, 24 vols., 2:301–22. London: Hogarth Press and Institute for Psycho-Analysis.

George, Alexander, and Daniel J. Velleman. 2002. *Philosophies of Mathematics*. Malden, Mass.: Blackwell.

Golding, Sue. 1995. "The Politics of Foucault's Poetics, or, Better Yet: The Ethical Demand of Ecstatic Fetish." In *Michel Foucault: J'Accuse (New Formations)*, ed. Judith Squires, 40–47. London: Lawrence & Wishart.

Grosz, Elizabeth. 1994. "A Thousand Tiny Sexes: Feminism and Rhizomatics." In *Gilles Deleuze and the Theater of Philosophy*, ed. Constantin V. Boundas and Dorothea Olkowski, 187–210. London: Routledge.

————. 2004. *The Nick of Time: Politics, Evolution, and the Untimely*. Durham: Duke University Press.

Habermas, Jürgen. 1987. *The Philosophical Discourse of Modernity*. Trans. Frederick Lawrence. Cambridge: Polity Press.

Hammond, Matthew. 2003. "How Does One Think with What Is Already Thinking?" Paper delivered at the Exeter SHiPSS Graduate Conference, 28 May, Exeter University, Exeter, UK.

Hardt, Michael. 1993. *Gilles Deleuze: An Apprenticeship in Philosophy*. London: University College London Press.

Hardt, Michael, and Antonio Negri. 2000. *Empire*. Cambridge: Harvard University Press.

Hartsock, Nancy. 1990. "Foucault and Power: A Theory for Women?" In *Feminism/Postmodernism*, ed. Linda Nicholson, 157–75. London: Routledge.

Hawkes, Terence. 1977. *Structuralism and Semiotics*. Berkeley and Los Angeles: University of California Press.

Hegel, G. W. F. 1970. *Philosophy of Nature: Being Part Two of the Encyclopaedia of the Philosophical Sciences*. Trans. A. V. Miller. Foreword by J. N. Findlay. Oxford: Oxford University Press.

———. 1975. *Hegel's Logic: Being Part One of the Encyclopaedia of the Philosophical Sciences*. Trans. William Wallace. Foreword by J. N. Findlay. Oxford: Oxford University Press.

———. 1977. *Phenomenology of Spirit*. Trans. A. V. Miller. Foreword by J. N. Findlay. Oxford: Oxford University Press.

Heidegger, Martin. 1962. *Being and Time*. Trans. John Macquarrie and Edward Robinson. Oxford: Blackwell.

———. 1969. *Identity and Difference*. Trans. Joan Stambaugh. New York: Harper & Row.

———. 1972. *On Time and Being*. Trans. Joan Stambaugh. New York: Harper & Row.

———. 1977. *The Question Concerning Technology and Other Essays*. Trans. William Lovitt. New York: Harper & Row.

———. 1979–87. *Nietzsche*. Ed. David Farrell Krell. 4 vols. San Francisco: HarperCollins.

———. 1982. *The Basic Problems of Phenomenology*. Rev. ed. Trans. Albert Hofstadter. Bloomington: Indiana University Press.

———. 1993. *Basic Writings*. Rev. and exp. ed. Ed. David Farrell Krell. San Francisco: HarperCollins.

Hintikka, Jaakko. 1979. "Aristotelian Infinity." In *Articles on Aristotle III: Metaphysics*, ed. Jonathan Barnes, Malcolm Schofield, and Richard Sorabji, 125–39. London: Gerald Duckworth & Co.

Houlgate, Stephen. 1986. *Hegel, Nietzsche, and the Criticism of Metaphysics*. Cambridge: Cambridge University Press.

Hyppolite, Jean. 1997. *Logic and Existence*. Trans. Leonard Lawlor and Amit Sen. Albany: State University of New York Press.

———. 2006. "A Spoken Commentary on Freud's 'Verneinung.'" In Jacques Lacan, *Écrits,* trans. Bruce Fink, in collaboration with Héloïse Fink and Russell Grigg, 746–54. New York: W. W. Norton.

Irigaray, Luce. 1985. *This Sex Which Is Not One*. Trans. Catherine Porter. Ithaca: Cornell University Press.

Jurist, Elliot L. 2000. *Beyond Hegel and Nietzsche: Philosophy, Culture, and Agency*. Cambridge: MIT Press.

Kahn, Charles H. 2007. "Why Is the *Sophist* a Sequel to the *Theaetetus?*" *Phronesis* 52 (1): 33–57.

Kant, Immanuel. 1965. *Critique of Pure Reason*. Unabridged ed. Trans. Norman Kemp Smith. New York: St. Martin's Press.

Kazarian, Edward. 1998. "Deleuze, Perversion, and Politics." *International Studies in Philosophy* 30 (1): 91–106.

Klein, Melanie. 1975. *Envy and Gratitude and Other Works: 1946–1963*. New York: Delacorte Press/Seymour Lawrence.

———. 1986. *The Selected Melanie Klein*. Ed. Juliet Mitchell. Harmondsworth: Penguin Books.

Kojève, Alexandre. 1969. *Introduction to the Reading of Hegel*. Assembled by Raymond Queneau, trans. James H. Nichols Jr., ed. Allan Bloom. New York: Basic Books.

Kristeva, Julia. 2001. *Melanie Klein*. Trans. Ross Guberman. New York: Columbia University Press.

Kuiken, Kir. 2005. "Deleuze/Derrida: Towards an Almost Imperceptible Difference." *Research in Phenomenology* 35 (1): 290–308.

Lacan, Jacques. 1977. *Écrits: A Selection*. Trans. Alan Sheridan. New York: W. W. Norton.

———. 1981. *The Four Fundamental Concepts of Psycho-Analysis*. Trans. Alan Sheridan. Ed. Jacques-Alain Miller. New York: W. W. Norton.

———. 1982. *Feminine Sexuality: Jacques Lacan and the École Freudienne*. Trans. Jacqueline Rose, ed. Juliet Mitchell and Jacqueline Rose. New York: W. W. Norton.

———. 2006. *Écrits*. Trans. Bruce Fink, in collaboration with Héloïse Fink and Russell Grigg. New York: W. W. Norton.

Laclau, Ernesto. 1990. *New Reflections on the Revolution of Our Time*. London: Verso.

Laclau, Ernesto, and Chantal Mouffe. 1985. *Hegemony and Socialist Strategy: Towards a Radical Democratic Politics*. London: Verso.

Lawlor, Leonard. 2002. *Derrida and Husserl: The Basic Problem of Phenomenology*. Bloomington: Indiana University Press.

———. 2003. *The Challenge of Bergsonism*. London: Continuum.

Lecercle, Jean-Jacques. 1989. *Philosophy Through the Looking-Glass: Language, Nonsense, Desire*. La Salle, Ill.: Open Court.

———. 2002. *Deleuze and Language*. Basingstoke, UK: Palgrave Macmillan.

Long, A. A. 1986. *Hellenistic Philosophy: Stoics, Epicureans, Sceptics*. 2d ed. London: Duckworth.

Long, A. A., and D. N. Sedley. 1987. *The Hellenistic Philosophers*. Vol. 1, *Translations of the Principal Sources with Philosophical Commentary*. Cambridge: Cambridge University Press.

Lorraine, Tasmin. 1999. *Irigaray and Deleuze: Experiments in Visceral Philosophy*. Ithaca: Cornell University Press.

MacNay, Lois. 1992. *Foucault and Feminism*. Cambridge, UK: Polity Press.

Malabou, Catherine. 1996. "Who's Afraid of Hegelian Wolves?" In *Deleuze: A Critical Reader*, ed. Paul Patton, 114–38. Oxford: Blackwell.

Mates, Benson. 1953. *Stoic Logic*. Berkeley and Los Angeles: University of California Press.

McTaggart, J. Ellis. 1908. "The Unreality of Time." *Mind: A Quarterly Review of Psychology and Philosophy* 17 (October): 457–74.

McWhorter, Ladelle. 1999. *Bodies and Pleasures: Foucault and the Politics of Sexual Normalization*. Bloomington: Indiana University Press.

Moulard, Valentine. 2002. "The Time-Image and Deleuze's Transcendental Empiricism." *Continental Philosophy Review* 35 (3): 325–45.

Mullarkey, John. 2005. "Forget the Virtual: Bergson, Actualism, and the Refraction of Reality." *Continental Philosophy Review* 37 (4): 469–93.

Murphy, Timothy S. 1999. "Beneath Relativity: Bergson and Bohm on Absolute Time." In *The New Bergsonism,* ed. John Mullarkey, 66–81. Manchester: Manchester University Press.

Newman, Saul. 2001. *From Bakunin to Lacan: Anti-Authoritarianism and the Dislocation of Power.* Lanham, Md.: Lexington Books.

Nietzsche, Friedrich. 1966. *Thus Spoke Zarathustra.* Trans. Walter Kaufmann. New York: Viking Press.

———. 1967. *On the Genealogy of Morals.* Trans. Walter Kaufmann and R. J. Hollingdale. New York: Vintage Books.

———. 1967–79. *Kritische Gesamtausgabe.* 30 vols. Ed. Giorgio Colli and Mazzino Montinari. Berlin: W. de Gruyter.

———. 1968. *The Will to Power.* Trans. Walter Kaufmann and R. J. Hollingdale. New York: Vintage Books.

———. 1974. *The Gay Science, with a Prelude in Rhymes and an Appendix of Songs.* Trans. Walter Kaufmann. New York: Vintage Books.

———. 1982. *Daybreak: Thoughts on the Prejudices of Morality.* Trans. R. J. Hollingdale. Cambridge: Cambridge University Press.

———. 1983. "On the Uses and Disadvantages of History for Life." In *Untimely Meditations,* trans. R. J. Hollingdale, 59–123. Cambridge: Cambridge University Press.

———. 1989. *Beyond Good and Evil: Prelude to a Philosophy of the Future.* Trans. Walter Kaufmann. New York: Vintage Books.

———. 1990. *Twilight of the Idols/The Anti-Christ.* Trans. R. J. Hollingdale. Harmondsworth: Penguin Books.

Nolan, Daniel. 2006. "Stoic Gunk." *Phronesis* 51 (2): 162–83.

Plato. 1961. *The Collected Dialogues of Plato, Including the Letters.* Ed. Edith Hamilton and Huntington Cairns. Princeton: Princeton University Press.

Pluth, Ed. 2006. "Lacan's Subversion of the Subject." *Continental Philosophy Review* 39 (3): 293–312.

Riemann, Bernhard. 1873. "On the Hypotheses Which Lie at the Bases of Geometry." Trans. W. K. Clifford. *Nature* 8 (183–84): 14–17, 36–37.

Rist, J. M. 1969. *Stoic Philosophy.* Cambridge: Cambridge University Press.

Robert, Paul. 1993. *Le nouveau petit Robert: Dictionnaire alphabétique et analogique de la langue francaise.* Paris: Le Robert.

Rose, Jaqueline. 1993. *Why War? Psychoanalysis, Politics and the Return to Melanie Klein.* Bucknell Lectures in Literary Theory. Oxford: Blackwell.

Russell, Bertrand. 1926. *Our Knowledge of the External World as a Field for Scientific Method in Philosophy.* London: George Allen & Unwin.

———. 1937. *The Principles of Mathematics.* 2d ed. London: George Allen & Unwin.

———. 1946. *History of Western Philosophy and Its Connection with Political and Social Circumstances from the Earliest Times to the Present Day.* London: George Allen and Unwin.

———. 1971. *Introduction to Mathematical Philosophy.* New York: Simon and Schuster.

Sadler, Ted. 1996. *Heidegger and Aristotle: The Question of Being*. London: Athlone Press.

Sambursky, S. 1959. *Physics of the Stoics*. London: Routledge and Kegan Paul.

Sartre, Jean-Paul. 1957. *The Transcendence of the Ego: An Existentialist Theory of Consciousness*. Trans. Forrest Williams and Robert Kirkpatrick. New York: Noonday Press.

Saussure, Ferdinand de. 1986. *Course in General Linguistics*. Trans. Roy Harris. Chicago: Open Court.

Scott, David. 2006. "The 'Concept of Time' and the 'Being of the Clock': Bergson, Einstein, Heidegger, and the Interrogation of the Temporality of Modernism." *Continental Philosophy Review* 39 (2): 183–213.

Shanker, S. G. 1987. *Wittgenstein and the Turning-Point in the Philosophy of Mathematics*. London: Croom Helm.

Smith, Daniel W. 1996. "Deleuze's Theory of Sensation: Overcoming the Kantian Duality." In *Deleuze: A Critical Reader*, ed. Paul Patton, 29–56. Oxford: Blackwell.

———. 1997. "Gilles Deleuze and the Philosophy of Difference: Toward a Transcendental Empiricism." Ph.D. diss., University of Chicago.

———. 2003. "Mathematics and the Theory of Multiplicities: Badiou and Deleuze Revisited." *Southern Journal of Philosophy* 41 (3): 411–49.

———. 2005. "The Concept of the Simulacrum: Deleuze and the Overturning of Platonism." *Continental Philosophy Review* 38 (1–2): 89–123.

Sorabji, Richard. 1983. *Time Creation and the Continuum: Theories in Antiquity and the Early Middle Ages*. London: Duckworth.

Stavrakakis, Yannis. 1999. *Lacan and the Political*. London: Routledge.

Tønder, Lars, and Lasse Thomassen, eds. 2005. *Radical Democracy: Politics Between Abundance and Lack*. Manchester: Manchester University Press.

Velleman, Daniel J. 1993. "Constructivism Liberalized." *Philosophical Review* 102 (1): 59–84.

Widder, Nathan. 2000. "What's Lacking in the Lack: A Comment on the Virtual." *Angelaki* 5 (3): 117–38.

———. 2001. "The Rights of Simulacra: Deleuze and the Univocity of Being." *Continental Philosophy Review* 34 (4): 437–53.

———. 2002. *Genealogies of Difference*. Urbana: University of Illinois Press.

———. 2004. "Foucault and Power Revisited." *European Journal of Political Theory* 3 (4): 411–32.

———. 2005. "Two Routes from Hegel." In *Radical Democracy: Politics Between Abundance and Lack,* ed. Lars Tønder and Lasse Thomassen, 32–49. Manchester: Manchester University Press.

Williams, James. 2003. *Gilles Deleuze's Difference and Repetition: A Critical Introduction and Guide*. Edinburgh: Edinburgh University Press.

———. 2005. "How Radical Is the New: Deleuze and Bachelard on the Problems of Completeness and Continuity in Dialectics." *Pli: The Warwick Journal of Philosophy* 16: 149–70.

Wittgenstein, Ludwig. 1975. *Philosophical Remarks*. Ed. Rush Rhees, trans. Raymond Hargreaves and Roger White. Oxford: Basil Blackwell.

———. 1978. *Remarks on the Foundations of Mathematics*. 3d ed. Ed. G. H. von Wright, R. Rhees, and G. E. M. Anscombe, trans. G. E. M. Anscombe. Oxford: Basil Blackwell.

Žižek, Slavoj. 1989. *The Sublime Object of Ideology*. London: Verso.

INDEX